# MILITARY LASER TECHNOLOGY FOR DEFENSE

# Military Laser Technology for Defense

## Technology for Revolutionizing 21st Century Warfare

**Alastair D. McAulay**
Lehigh University, Bethlehem, PA

**WILEY**

A JOHN WILEY & SONS, INC., PUBLICATION

Published by John Wiley & Sons, Inc., Hoboken, New Jersey
Published simultaneously in Canada

For general information on our other products and services or for technical support, please contact our Customer Care Department within the United States at (800) 762-2974, outside the United States at (317) 572-3993 or fax (317) 572-4002.

Wiley also publishes its books in a variety of electronic formats. Some content that appears in print may not be available in electronic formats. For more information about Wiley products, visit our web site at www.wiley.com.

ISBN 978-0-470-25560-5

*Library of Congress Cataloging-in-Publication Data is available.*

10 9 8 7 6 5 4 3 2 1

# CONTENTS

# PREFACE

In 1832, Carl Von Clausewitz [22] wrote: "War is an extension of politics." Historically, war erupts when groups cannot resolve their conflicts politically. Consequently, every group must prepare to defend itself against reasonable future threats.

Laser technology is ideal for defense against modern weapons because laser beams can project energy over kilometers in microseconds, fast enough to eliminate most countermeasure responses. This book includes only unclassified or declassified information and focuses on military applications that involve propagation through the atmosphere. Chapters 1–6 provide background material on optical technologies. Chapters 7–11 describe laser technologies including efficient ultrahigh-power lasers such as the free-electron laser that will have a major impact on future warfare. Chapters 12–17 show how laser technologies can effectively mitigate six of the most pressing military threats of the 21st century. This includes the use of lasers to protect against missiles, future nuclear weapons, directed beam weapons, chemical and biological attacks, and terrorists and to overcome the difficulty of imaging in bad weather conditions.

Understanding these threats and their associated laser protection systems is critical for allocating resources wisely because a balance is required between maintaining a strong economy, an effective infrastructure, and a capable military defense. A strong defense discourages attackers and is often, in the long run, more cost-effective than alternatives. I believe laser technology will revolutionize warfare in the 21st century.

ALASTAIR D. MCAULAY
Lehigh University, Bethlehem, PA

# ACKNOWLEDGMENTS

I thank my wife Carol-Julia, for her patience and help with this book. Also my thanks to my son Alexander and his wife Elizabeth. I wish to acknowledge the too-many-to-name researchers in this field whose publications I have referenced or with whom I have had discussions. This includes appreciation for the International Society for Optical Engineering (SPIE), the Optical Society of America (OSA), and IEEE. I also thank Lehigh University for providing me with the environment to write this book.

# ABOUT THE AUTHOR

Alastair McAulay received a PhD in Electrical Engineering from Carnegie Mellon University, and an MA and BA in Mechanical Sciences from Cambridge University. Since 1992, he is a Professor in the Electrical and Computer Engineering Department at Lehigh University; he was Chandler-Weaver Professor and Chair of EECS at Lehigh from 1992 to 1997 and NCR Distinguished Professor and Chair of CSE at Wright State University from 1987 to 1992. Prior to that, he was in the Corporate Laboratories of Texas Instruments for 8 years, where he was program manager for a DARPA optical data flow computer described in his book "Optical Computer Architectures" that was published by Wiley in 1991. Prior to that, he worked in the defense industry on projects such as the Advanced Light Weight Torpedo that became the Mk. 50 torpedo. Dr McAulay can be contacted via network Linked In.

**PART I**

---

# OPTICS TECHNOLOGY FOR
# DEFENSE SYSTEMS

---

# CHAPTER 1

# OPTICAL RAYS

Geometric or ray optics [16] is used to describe the path of light in free space in which propagation distance is much greater than the wavelength of the light—normally microns (see Section 1.2.3 for more exact conditions). Note that we cannot apply ray theory if the media properties vary noticeably in distances comparable to wavelength; for such cases, we use more computationally demanding finite approximation techniques such as finite-difference time domain (FDTD) [154] or finite elements [78, 79]. Ray theory postulates rays that are at right angles to wave fronts of constant phase. Such rays describe the path along which light emanates from a source and the rays track the Poynting vector of power in the wave. Geometric or ray optics provides insight into the distribution of energy in space with time. The spread of neighboring rays with time enables computation of attenuation, which provides information analogous to that provided by diffraction equations but with less computation. Ray optics is extensively used for the passage of light through optical elements, such as lenses, and inhomogeneous media for which refractive index (or dielectric constant) varies with position in space.

In Section 1.1, we derive the paraxial equation that reduces dimensionality when light stays close to the axis. In Section 1.2, we study geometric or ray optics: Fermat's principle, limits of ray theory, the ray equation, rays through quadratic media, and matrix representations. In Section 1.3, we consider thin lens optics for launching and/or receiving beams: magnification, beam expanders, beam compressors, telescopes, microscopes, and spatial filters.

---

*Military Laser Technology for Defense: Technology for Revolutionizing 21st Century Warfare*,
First Edition. By Alastair D. McAulay.
© 2011 John Wiley & Sons, Inc. Published 2011 by John Wiley & Sons, Inc.

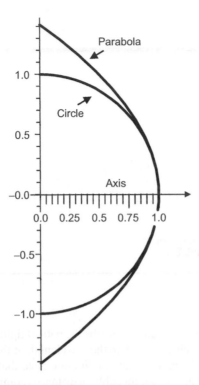

**FIGURE 1.1** Illustrates the paraxial approximation.

## 1.1 PARAXIAL OPTICS

In 1840, Gauss proposed the paraxial approximation for propagation of beams that stay close to the axis of an optical system. In this case, propagation is, say, in the $z$ direction and the light varies in transverse $x$ and $y$ directions over only a small distance relative to the distance associated with the radius of curvature of a spherically curved surface in $x$ and $y$ (Figure 1.1). The region of the spherical surface near the axis can be approximated by a parabola. The spherical surface of curvature $R$ is

$$x^2 + y^2 + z^2 = R^2 \quad \text{or} \quad z = R\sqrt{\left(1 - \frac{x^2 + y^2}{R^2}\right)} \tag{1.1}$$

Using the binomial theorem to eliminate the square root,

$$z = R\left(1 - \frac{x^2 + y^2}{2R^2}\right) \quad \text{or} \quad R - z = \frac{x^2 + y^2}{2R} \tag{1.2}$$

which is the equation for a parabola.

## 1.2  GEOMETRIC OR RAY OPTICS

### 1.2.1  Fermat's Principle

In 1658, Fermat introduced one of the first variational principles in physics, the basic principle that governs geometrical optics [16]: A ray of light will travel between points $P_1$ and $P_2$ by the shortest optical path $L = \int_{P_1}^{P_2} n \, ds$; no other path will have a shorter optical path length. The optical path length is the equivalent path length in air for a path through a medium of refractive index $n$. Equivalently, because the refractive index is $n = c/v$ ($v$ is the phase velocity, and $c$ is the velocity of light), $n \, ds = c \, dt$, this is also the shortest time path. As the optical path length or time differs for each path, our optimization to determine the shortest (a minimum extremum) is that of a length or time function among many path functions, that is a function of a function (a functional), and this requires the use of calculus of variations [42]. *Fermat's principle* is written for minimum optical path length or, equivalently, for minimum time:

$$\delta L = \delta \int_{P_1}^{P_2} n \, ds = 0 \quad \text{or} \quad \delta L = \delta \int_{P_1}^{P_2} c \, dt = 0 \tag{1.3}$$

Fermat's principle lends itself to geometric optics in which light is considered to be rays that propagate at right angles to the phase front of a wave, normally in the direction of the Poynting power vector. Note that electromagnetic waves are transverse, and the electric and magnetic fields in free space oscillate at right angles to the direction of propagation and hence to the ray path. When valid, a wave can be represented more simply by a single ray.

### 1.2.2  Fermat's Principle Proves Snell's Law for Refraction

Fermat's principle can be used to directly solve problems of geometric optics as illustrated by our proof of Snell's law of refraction, the bending at an interface between two media of different refractive indices $n_1 = \sqrt{\mu_1 \epsilon_1}$ and $n_2 = \sqrt{\mu_2 \epsilon_2}$, where $\epsilon$ is the dielectric constant and $\mu$ is the relative permeability (Figure 1.2). From Fermat's principle, the optical path from $P_1$ to $P_2$ intercepts the dielectric interface at $R$ so that the optical path length through $R$ is the least for all possible intercepts at the interface. Because at an extremum the function in equation (1.3) has zero gradient, moving the intercept point a very small variational distance $\delta x$ along the interface to $Q$ will not change the optical path length. From Figure 1.2, the change in optical path length when moving from the path through $R$ to the path through $Q$ is

$$\delta s - \delta s' = n \delta x \sin \theta - n' \delta x \sin \theta' = 0 \tag{1.4}$$

which gives Snell's law

$$n \sin \theta = n' \sin \theta' \tag{1.5}$$

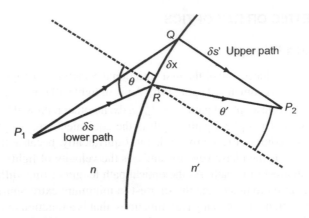

**FIGURE 1.2**   Deriving Snell's law from Fermat's principle.

When light passes through an inhomogeneous medium for which refractive index $n(\mathbf{r}) = n(x, y, z)$ varies with position, a ray will no longer be straight. The divergence of adjacent rays provides an estimate of the attenuation as a function of distance along the ray.

### 1.2.3   Limits of Geometric Optics or Ray Theory

Rays provide an accurate solution to the wave equation only when the radius of curvature of the rays and the electric field vary only slowly relative to wavelength, which is often the case for light whose wavelength is only in micrometers. When light rays come together as in the focus region from a convex lens, rapid changes in field can occur in distances comparable to a wavelength. Hence, rays are inaccurate representations for the solution to the wave equation at so-called caustics (the envelope formed by the intersection of adjacent rays).

As light travels straight in a constant medium, we can discretize our region into small regions of different but fixed refractive index in a finite-difference technique. The rays across a small region are then coupled into adjacent regions using Snell's law. The steps in refractive index between small regions can cause spurious caustics in the ray diagrams, which can be minimized by switching to a piecewise linear refractive index approximation, as in a first-order polynomial finite-element approach [78, 79]. To plot a ray from a source to a target, we can draw multiple rays starting out at difference angles from the source until we find one that passes through the target; this is a two-point boundary problem.

### 1.2.4   Fermat's Principle Derives Ray Equation

The ray equation, critical in geometric optics, describes the path of an optical ray through an inhomogeneous medium in which refractive index changes in 3D space [16, 148, 176]. The optical path length for use in Fermat's principle, equation

(1.3), may be written by factoring out $dz$ from $ds = \sqrt{dx^2 + dy^2 + dz^2}$:

$$\delta \int_{P_1}^{P_2} n \, ds = \delta \int_{z_1}^{z_2} n(x, y, z) \sqrt{\left(\frac{dx}{dz}\right)^2 + \left(\frac{dy}{dz}\right)^2 + 1} \, dz$$

$$= \delta \int_{z_1}^{z_2} n(x, y, z) \sqrt{x'^2 + y'^2 + 1} \, dz \qquad (1.6)$$

where prime indicates $d/dz$ and $ds = \sqrt{x'^2 + y'^2 + 1} \, dz$. Equation (1.6) can be written as $\delta \int_{z_1}^{z_2} F \, dz$, where the integrand $F$ has the form of a functional (function of functions)

$$F(x', y', x, y, z) \equiv n(x, y, z) \sqrt{x'^2 + y'^2 + 1} \qquad (1.7)$$

From calculus of variations [16], the solutions for extrema (maximum or minimum) with integrand of the form of equation (1.7) are the Euler equations

$$F_x - \frac{d}{dz} F_{x'} = 0, \qquad F_y - \frac{d}{dz} F_{y'} = 0 \qquad (1.8)$$

where subscripts refer to partial derivatives. From equation (1.7) and $x' = dx/dz$,

$$F_x = \frac{\partial n}{\partial x} ds = \frac{\partial n}{\partial x} \sqrt{x'^2 + y'^2 + 1} = \frac{\partial n}{\partial x} \frac{ds}{dz} \qquad (1.9)$$

and

$$F_{x'} = n \frac{1}{2\sqrt{x'^2 + y'^2 + 1}} 2x' = n \frac{dx}{dz} \frac{dz}{ds} = n \frac{dx}{ds} \qquad (1.10)$$

Similar equations apply for $F_y$ and $F_{y'}$. Substituting equations (1.9) and (1.10) into equation (1.8) gives

$$\frac{\partial n}{\partial x} \frac{ds}{dz} - \frac{d}{dz} \left( n \frac{dx}{ds} \right) = \frac{\partial n}{\partial x} - \frac{dz}{ds} \frac{d}{dz} \left( n \frac{dx}{ds} \right) = 0 \qquad (1.11)$$

The resulting equations for the ray path are

$$\frac{d}{ds} \left( n \frac{dx}{ds} \right) = \frac{\partial n}{\partial x}, \qquad \frac{d}{ds} \left( n \frac{dy}{ds} \right) = \frac{\partial n}{\partial y}, \qquad \frac{d}{ds} \left( n \frac{dz}{ds} \right) = \frac{\partial n}{\partial z} \qquad (1.12)$$

where the last equation is obtained by reassigning coordinates, by analogy, or by additional algebraic manipulation [16]. These equations can be written in vector form for the *vector ray equation*

$$\frac{d}{ds}\left(n\frac{dr}{ds}\right) = \nabla n \tag{1.13}$$

Another derivation [16] for the ray equation provides a different perspective. The derivation generates, from Maxwell's equations or from the wave equation, an equivalent to Fermat's principle, *the eikonal equation*.

$$(\nabla S)^2 = n^2 \quad \text{or} \quad \left(\frac{\partial S}{\partial x}\right)^2 + \left(\frac{\partial S}{\partial y}\right)^2 + \left(\frac{\partial S}{\partial z}\right)^2 = n^2(x, y, z) \tag{1.14}$$

The eikonal equation relates phase fronts $S(\mathbf{r}) = $ constant and refractive index $n$. A ray $n\mathbf{s}$ is in the direction at right angles to the phase front, that is, in the direction of the gradient of $S(\mathbf{r})$, or

$$n\mathbf{s} = \nabla S \quad \text{or} \quad n\frac{d\mathbf{r}}{ds} = \nabla S \tag{1.15}$$

By taking the derivative of equation (1.15) with respect to $s$, we obtain the ray equation (1.13).

### 1.2.5 Useful Applications of the Ray Equation

We illustrate the ray equation for rays propagating in a $z$–$y$ plane of a slab, where $z$ is the propagation direction axis for the paraxial approximation and refractive index varies transversely in $y$. For a homogeneous medium, $n$ is constant and $\nabla n = \partial n/\partial y = 0$. Then the ray equation (1.13) becomes $d^2 y/dz^2 = 0$. After integrating twice, $y = az + b$, a straight line in the $z$–$y$ plane. Therefore, in a numerical computation, we discretize the refractive index profile into piecewise constant segments in $y$ and obtain a piecewise linear optical ray path in plane $z$–$y$.

For a linearly varying refractive index in $y$, $n = n_0 + ay$ with $n \approx n_0$, $\partial n/\partial y = a$, the ray equation (1.13) becomes $d^2 y/dz^2 \approx a/n_0$. After two integrations, $y = (a/n_0)z^2 + (b/n_0)z + d$, which is a quadratic in the $z$–$y$ plane and can be represented to first approximation by a spherical arc. Therefore, if we discretize the refractive index profile into piecewise linear segments, we obtain a ray path of joined arcs that is smoother than the piecewise linear optical ray path for a piecewise constant refractive index profile. The approach is extrapolatable to higher dimensions.

Another useful refractive index profile is that of a quadratic index medium, in which the refractive index smoothly decreases radially out from the axis of a cylindrical body (Figure 1.3a):

$$n^2 = n_0^2(1 - (gr)^2) \quad \text{with} \quad r^2 = (x^2 + y^2) \tag{1.16}$$

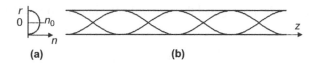

**FIGURE 1.3** Ray in quadratic index material: (a) refractive index profile and (b) ray path.

where $g$ is the strength of the curvature and $gr \ll 1$. Material doping creates such a profile in graded index fiber to replace step index fiber. In a cylindrical piece of glass, such an index profile will act as a lens, called a GRIN lens [47]. A GRIN lens can be attached to the end of an optical fiber and can match the fiber diameter to focus or otherwise image out of the fiber. In the ray equation, for the paraxial approximation, $d/ds = d/dz$, and from equation (1.16), $\partial n/\partial r = -n_0 g^2 r$. So the ray equation reduces to

$$\frac{d^2 r}{dz^2} + g^2 r = 0 \tag{1.17}$$

which has sin and cos solutions. A solution to equation (1.17) with initial conditions $(r_0)_{in}$ and $(dr_0/dz)_{in} = (r_0')_{in}$ is

$$r = (r_0)_{in} \cos(gz) + (r_0')_{in} \frac{\sin(gz)}{g} \tag{1.18}$$

which can be verified by substituting into equation (1.17). A ray according to equation (1.18) for a profile, equation (1.17), is shown in Figure 1.3b.

### 1.2.6 Matrix Representation for Geometric Optics

The ability to describe paraxial approximation propagation in the $z$ direction through circularly symmetric optical components using a location and a slope in geometric optics allows for a $2 \times 2$ matrix representation [44, 132, 176].

We consider a material of constant refractive index and width $d$. For this medium, light propagates in a straight line (Section 1.2.3), and a ray path does not change slope, $r_{out}' = r_{in}'$. The ray location changes after passing through width $d$ of this medium according to

$$r(z)_{out} = r(z)_{in} + r'(z)_{in} d \tag{1.19}$$

where after traveling a distance $d$ at slope $r'$, location has changed by $r'(z)_{in} d$.

Hence, we can write a matrix equation relating the output and the input for a position and a slope vector $[r(z), r'(z)]^T$:

$$\begin{bmatrix} r(z) \\ r'(z) \end{bmatrix}_{out} = \begin{bmatrix} 1 & d \\ 0 & 1 \end{bmatrix} \begin{bmatrix} r(z) \\ r'(z) \end{bmatrix}_{in} \tag{1.20}$$

Similarly, the ray can be propagated through a change in refractive index from $n_1$ to $n_2$ with

$$
\begin{bmatrix} r(z) \\ r'(z) \end{bmatrix}_{\text{out}} = \begin{bmatrix} 1 & 0 \\ 0 & \frac{n_1}{n_2} \end{bmatrix} \begin{bmatrix} r(z) \\ r'(z) \end{bmatrix}_{\text{in}}
\tag{1.21}
$$

where position does not change and from Snell's law for small angles, for which slope $\tan\theta \approx \sin\theta$, the slope changes by $n_1/n_2$.

Another common matrix is that for passing through a lens of focal length $f$:

$$
\begin{bmatrix} r(z) \\ r'(z) \end{bmatrix}_{\text{out}} = \begin{bmatrix} 1 & 0 \\ \frac{1}{-f} & 1 \end{bmatrix} \begin{bmatrix} r(z) \\ r'(z) \end{bmatrix}_{\text{in}}
\tag{1.22}
$$

where a lens changes the slope of a ray by $-r(z)/f$.

A useful case is the propagation of rays through a quadratic medium. From equation (1.18), a $2 \times 2$ matrix can be written and verified by substituting

$$
\begin{bmatrix} r(z) \\ r'(z) \end{bmatrix}_{\text{out}} = \begin{bmatrix} \cos(gz) & \frac{\sin(gz)}{g} \\ -g\sin(gz) & \cos(gz) \end{bmatrix} \begin{bmatrix} r(z) \\ r'(z) \end{bmatrix}_{\text{in}}
\tag{1.23}
$$

Other matrices are illustrated in Ref. [176]. The advantage of the $2 \times 2$ representation is that for a string (or sequence) of circularly symmetric components, the matrices can be multiplied together to achieve a single $2 \times 2$ matrix for transmission through the complete string. The property is that the determinant of any matrix is zero. We will use in Section 2.1.2 the $2 \times 2$ notation with matrix elements labeled clockwise from top left as ABCD to compute the effect of propagating a Gaussian beam through the corresponding optical element.

## 1.3  OPTICS FOR LAUNCHING AND RECEIVING BEAMS

Ray tracing allows modeling of simple optics for launching and receiving beams. Beam expanders, beam compressors, telescopes, microscopes, and spatial filters are frequently used in military optical systems to change the beam diameter, view an object at different levels of magnification, or improve the beam spatial coherence. These systems can be constructed with two thin refractive lenses [61]. More complex lens designs can be performed with commercial software such as Code V. A single thin lens system and a magnifier are discussed first.

### 1.3.1  Imaging with a Single Thin Lens

*1.3.1.1  Convex Lens for Imaging*    The focal length of a convex (positive) lens is the distance $f'$ at which parallel rays (a collimated beam) are focused to a point $F'$, (Figure 1.4a) [61]. A single lens can be used for imaging, that is, to create a copy

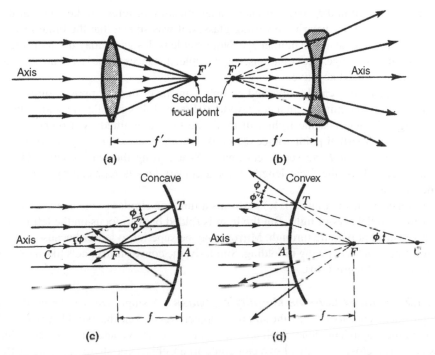

**FIGURE 1.4** Focusing a collimated parallel beam: (a) with a convex lens, (b) with a concave lens, (c) with a concave mirror, and (d) with a convex mirror.

of an input object to an output image of different size and location (Figure 1.5). An object $U_0$ is at distance $d_o$ (o for object) in front of the lens $L$ of focal length $f$. A copy, called the image $U_i$ (i for image), is located at a distance $d_i$ behind the lens (o and i are not to be confused with output and input). For a sharp image, the lens equation must be satisfied.

$$\frac{1}{d_o} + \frac{1}{d_i} = \frac{1}{f} \quad \text{and} \quad m = \frac{-d_i}{d_o} \tag{1.24}$$

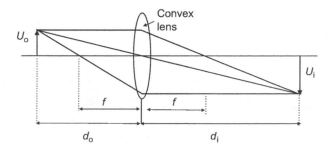

**FIGURE 1.5** Imaging with a single thin lens.

The negative sign in the image or lateral magnification $m$ refers to the fact that the image is inverted. By wearing inverting glasses, it was shown that the brain inverts the image in the case of the human eye. Note that lens designers may use a different convention that changes the equations; for example, distances to the left of an element are often considered negative.

Note that figures can be reversed as light can travel in the opposite directions through lenses and mirrors. In a concave lens (Figure 1.4b), parallel rays are caused to diverge. A viewer at the right will think the light is emitted by a point source at $F'$. This is a virtual point source as, unlike with a convex lens, a piece of paper cannot be placed at $F'$ to see a real image. When using the lens equation (1.24) for a concave lens, the focal length $f$ for a convex lens is replaced by $-f$ for a concave lens.

A concave mirror (Figure 1.4c) performs a function similar to the convex lens in focusing parallel rays of light. But the light is folded back to focus on the left of the mirror instead of passing through. Mirrors may be superior to lenses because of less weight and small size owing to folding. Similarly, the convex mirror acts like a folded concave lens (Figure 1.4d).

### 1.3.1.2 Convex Lens as Magnifying Glass
A single lens can be used as a simple microscope to increase the size of an object over the one that would be obtained without the magnifying lens. Such a system is used as an eyepiece in more complex systems. The closest a typical eye can come to an object for sharp focusing is the standard distinct image distance of $s' = 25$ cm. If it were possible to see an object closer to the eye, the image would occupy a larger area of the retina and the object would look larger. The magnifying glass allows the object to be brought closer than the minimum sharp distance of the eye, say to a distance $d_0$ in front of the eye, by projecting a virtual image at the standard distance, $s'$ (Figure 1.6) [61]. From the lens law, equation (1.24), using a negative sign for $s'$ because it is on the opposite side of the lens relative to Figure 1.5,

$$\frac{1}{d_0} = \frac{1}{s'} + \frac{1}{f} = \frac{f + s'}{fs'} \tag{1.25}$$

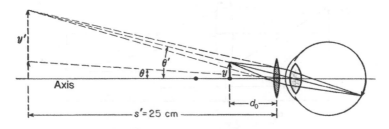

**FIGURE 1.6** A magnifying glass.

The angle $\theta$ subtended by the object in the absence of a magnifying lens and the angle $\theta'$ subtended with the magnifying glass are

$$\tan \theta = \frac{y}{s'}$$

$$\tan \theta' = \frac{y}{d_o} = y\frac{f + s'}{fs'} \tag{1.26}$$

where the second equation used equation (1.25). Therefore, the angular or power magnification may be written for small angles, using equation (1.25) for $1/d_o$, as

$$M = \frac{\theta'}{\theta} = \frac{s'}{d_o} = \frac{s'}{f} + 1 \approx \frac{s'}{f} \tag{1.27}$$

For $f$ in centimeters, and minimum distinct distance of $s' = 25$ cm, magnifying power is $M = 25/f$. An upper case $M$ distinguishes from lateral magnification $m$ in equation (1.24).

### 1.3.2 Beam Expanders

Beam expanders are used to increase beam diameter for beam weapons and optical communications. Beam expansion reduces the effects of diffraction when propagating light through the atmosphere. A source with a larger beam diameter will spread less with distance than one with a smaller diameter (Section 3.3.5) or, for example, if $\Delta s$ in equation (3.20) increases, then $\Delta \theta$ decreases (Section 3.2.2). Therefore, when a beam is launched into the air for optical communications, to replace a microwave link, or for a power ray, the beam diameter is expanded to minimize beam spreading. A wider beam is less influenced by turbulence because of averaging across the beam (Chapter 5).

Figure 1.7 shows how two convex lenses $L_1$ and $L_2$ of different focal lengths, $f_1$ and $f_2$, can expand a collimated beam diameter from $d_1$ to $d_2$. By similar triangles,

$$\frac{d_2}{d_1} = \frac{f_2}{f_1} \tag{1.28}$$

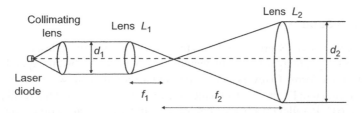

**FIGURE 1.7** Beam expander to reduce effects of beam spreading in the atmosphere.

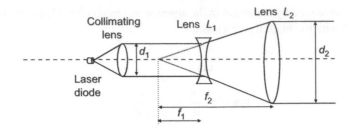

**FIGURE 1.8**  Beam expander made with a concave lens as the first lens.

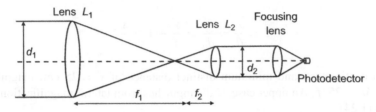

**FIGURE 1.9**  Beam compressor to reduce collimated beam diameter.

A beam expander can also be made shorter by using a concave lens for the first lens $L_1$ (Figure 1.8).

### 1.3.3  Beam Compressors

A compressor is the reverse of the expander as shown in Figure 1.9. In a receiver for an optical communication link, an incoming collimated beam is reduced in diameter from $d_1$ to $d_2$ to match the size of an optical sensor. By similar triangles, the image or lateral magnification, $m$ is

$$m = \frac{d_2}{d_1} = \frac{f_2}{f_1} \tag{1.29}$$

In a practical optical link, the beam expander on the transmit side forms a slightly converging beam. As the beam profile is normally Gaussian (Section 2.1), the propagation follows that described in Section 2.1.2.

### 1.3.4  Telescopes

The beam compressor (Section 1.3.3) has the form of a refractive telescope $L_1$ forms an image and $L_2$ reimages to $\infty$ for viewing by eye and the beam expander (Section 1.3.2) has the form of a reverse telescope. A more common drawing for a refractive telescope is shown in Figure 1.10, in which the real image at $Q'$ is at the focal point of both lenses [61]. Parallel rays from an object at infinity arrive at an object field angle $\theta$ to the axis and form an image at $Q'$. $2\theta$ is called the field of view. The objective lens

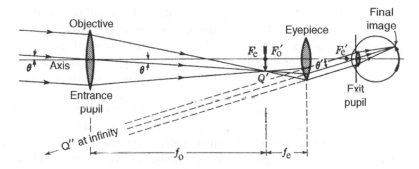

**FIGURE 1.10** Telescope.

acts as the aperture stop or the entrance pupil in the absence of a separate stop [61]. The second lens, usually called an eyepiece, magnifies the image at $Q'$ so that a larger virtual image $Q''$ appears at infinity (Section 1.3.1.2). The virtual image subtends an angle $\theta'$ at the eye. Angular magnification or magnifying power $M$ (reciprocal of lateral magnification) is

$$M = \frac{\theta'}{\theta} = \frac{f_\mathrm{o}}{f_\mathrm{e}} \tag{1.30}$$

Large astronomical telescopes built with refractive lenses are limited to approximately 1 m diameter because of the weight of the lenses. Higher resolution telescopes with larger diameters use mirrors and are discussed next.

***1.3.4.1 Cassegrain Telescope*** The Cassegrain telescope has a common dish appearance and is used in military systems to reduce weight and size relative to a refractive lens telescope for transmitting and receiving signals. (see Sections 16.2.5, 15.1.1 and 12.2). Figure 1.11a shows the inverted telescope as a beam expander for

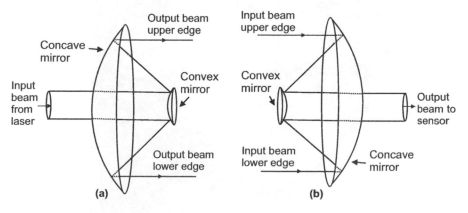

**FIGURE 1.11** Cassegrain antenna as (a) beam expander or inverted telescope and (b) telescope.

transmitting light beams. The input beam passes through a small hole in the large concave mirror to strike the small convex mirror. Comparing the Cassegrain inverted telescope with the lens beam expander in Figure 1.8, the first small concave lens, $L_1$, is replaced by a small convex mirror that spreads the light over a concave mirror that replaces the second lens $L_2$. The output aperture size is close to that of the large mirror diameter.

The reverse structure acts as a telescope (Figure 1.11b). The large concave mirror aperture determines the resolution of images. Light reflecting from the concave mirror focuses on the small convex mirror and then through a hole in the concave mirror onto a CCD image sensor. This configuration is used in the Geoeye imaging satellite 400 miles up (Figure 1.12) [125]. Such imaging satellites are critical for providing intelligence information for the military and data for commercial ventures such as Google. A Geoeye satellite, launched in 2008, as shown in Figure 1.13 [125], involves many other systems, solar panels, global positioning system (GPS), star tracker (together the star tracker and the GPS can locate objects to within 3 m), image storage, and data antenna for transmitting signals back to earth when over designated ground stations. In the open literature as of 2009, there are in orbit 51 imaging satellites with resolution between 0.4 and 56 m launched by 31 countries and 10 radar satellites launched by 18 countries [125]. The military and commercial sectors rely on these and classified satellites for intelligence relating to threat warnings of enemy activities and environmental issues, on global positioning satellites for guiding missiles and locating U.S. and allies personnel and vehicles, on communication satellites for battlefield communications,

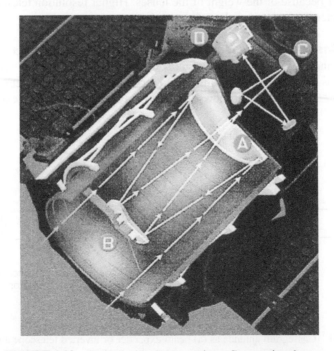

**FIGURE 1.12**   Optics inside Geoeye using a Cassegrain telescope.

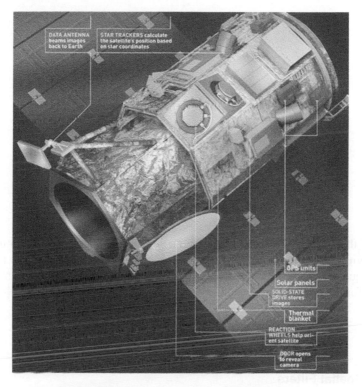

**FIGURE 1.13** Geoeye imaging satellite.

and on classified antisatellite satellites aimed at interfering with other countries' satellites. Hence, the control of satellite space will be critical in future wars, although in recent wars control of air space was adequate. Most satellites are vulnerable to laser attack from the ground, aircraft, or other satellites. For example, imaging satellites can be blinded by glare from lasers and for most satellites the solar cell arrays can be easily damaged by lasers, which can disable their source of solar energy. Consequently, as discussed in Chapter 14, the military satellites should also have laser warning devices and protection such as their own lasers and electronic countermeasures.

***1.3.4.2 Nasmyth Telescope*** Sometimes for convenience of mounting subsequent equipment, such as optical spectral analyzers, a variation of the Cassegrain telescope is used in which the light is brought out to one side using a third mirror, rather than through a hole in the primary mirror. This is referred to as a Nasmyth telescope (related to a Coudé telescope). Such an arrangement is shown diagrammatically in Figure 15.1.

### 1.3.5 Microscopes

A typical two-lens microscope (Figure 1.14) has a form similar to the beam expander (Section 1.3.2). A tiny object, in this case an arrow, is placed just inside the focal

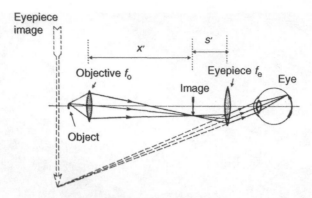

**FIGURE 1.14**   Microscope.

length of the objective lens. According to the lens law, equation (1.24), an image is formed with magnification $m_o = x'/f_o$. The eyepiece focal length $f_e$ has magnification, equation (1.27) (Section 1.3.1.2), $M_e = s'/f_e$, where the minimum distinct distance for the eye is $s' = 25$ cm. Consequently, the magnification is [61]

$$M = m_o M_e = \frac{x'}{f_o} \frac{s'}{f_e} \tag{1.31}$$

### 1.3.6   Spatial Filters

Spatial filters are used to improve spatial coherence in interferometers and between power amplifier stages in a high-power laser (Chapter 8 and Section 13.2.1). A spatial amplifier looks like Figure 1.7 but has a pinhole of very small size, usually micrometers, placed exactly at the focal point of the two lenses (Figure 1.15). Alignment of the pinhole requires high precision to make sure the pinhole lines up exactly with the main power at the focus of the beams. The light to the right of the pinhole now appears to come from an almost perfect point source that produces an almost perfect spherical wave. The smaller the pinhole, the closer is the wave to perfectly spherical. Note that light will be lost if the pinhole is too small. A collimating lens following the pinhole as in Figure 1.7 converts the spherical wave into an almost perfect plane wave.

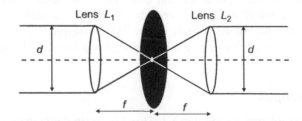

**FIGURE 1.15**   Spatial filter improves spatial coherence for higher quality beams.

Optical amplifiers cause distortion due to nonlinear effects at high power. Spatial correlation is degraded and the plane wave has regions pointing off axis. Hence, a spatial filter is often used to clean up the beam after amplification. In a series of power amplifiers (Section 13.2.1), a spatial filter after each amplifier will prevent distortion building up to unacceptable levels. The beam can even be passed back and forth through the same amplifier, as in the National Infrastructure Laser (Chapter 13), because the flash light duration is long enough for several passes of the beam to be amplified. An alternative method of improving beam quality by adaptive optics is described in Section 5.3.2 and used in the airborne laser in Section 12.2.2 and 12.2.3.

# CHAPTER 2

# GAUSSIAN BEAMS AND POLARIZATION

Laser beams are generally brighter at the center and have a cross-sectional intensity close to a Gaussian or a normal distribution. They also tend to have a defined polarization or direction of motion of the electric field. For example, in a laser diode with a rectangular waveguide, the polarization may be in the direction of the longer side of the rectangle.

In Section 2.1, we analyze Gaussian beams, characterized by spot size and curvature, and their propagation through lens systems. In section 2.2, we describe how to characterize, analyze, and control polarization.

## 2.1 GAUSSIAN BEAMS

The output of a laser is generally brighter at the center than at the edges and the cross-sectional intensity profile can often be approximated with a Gaussian distribution. This forms a Gaussian beam [148, 176]. The type of laser influences how well the far field from the laser satisfies the approximation. In a laser diode (Section 7.2), the beam emits from a small waveguide, which by diffraction causes the beam to spread widely. Coherent light emitted from a smaller diameter source will spread more with distance than light from a larger diameter source (Section 3.3.5) or, for example, if $\Delta s$ in equation (3.20) decreases, then the spreading angle $\Delta\theta$ increases (Section 3.2.2). A lens brings it back to a narrower beam pattern with a close approximation to Gaussian.

*Military Laser Technology for Defense: Technology for Revolutionizing 21st Century Warfare*, First Edition. By Alastair D. McAulay.
© 2011 John Wiley & Sons, Inc. Published 2011 by John Wiley & Sons, Inc.

In other lasers, concave curved mirrors reinforce the formation of a Gaussian beam [176].

Analogous to specifying Gaussian probability density functions with only two parameters (mean and variance), Gaussian beams can also be specified with only two parameters: in this case, radius of curvature $R$ and spot radius $W$. By tracking these two parameters, we can analyze propagation of Gaussian beams through optical elements and turbulence.

## 2.1.1 Description of Gaussian Beams

The complex amplitude of a plane wave propagating in the $z$ direction is modulated with a complex *envelope* $A(\mathbf{r})$ that has a Gaussian shape transversely and varies with $z$ as the beam expands and contracts:

$$U(\mathbf{r}) = A(\mathbf{r}, z)\exp\{-jkz\} \tag{2.1}$$

where we ignored polarization effects for simplicity.

The modulated wave at a single laser frequency in air satisfies the time harmonic or Helmholtz wave equation

$$\nabla^2 U + k^2 U = 0 \tag{2.2}$$

We separate out the transverse component in $\mathbf{r}$, where we assumed cylindrical symmetry, and the component in the direction of propagation $z$ to obtain

$$\frac{\partial^2}{\partial \mathbf{r}^2}U + \frac{\partial^2}{\partial z^2}U + k^2 U = 0 \tag{2.3}$$

We now derive equation (2.7) for the envelope $A(\mathbf{r}, z)$ by substituting the field from equation (2.1) into equation (2.3)

$$\frac{\partial^2}{\partial \mathbf{r}^2}(A(\mathbf{r}, z))\exp\{-jkz\}) + \frac{\partial^2}{\partial z^2}(A(\mathbf{r}, z)\exp\{-jkz\}) + k^2 A(\mathbf{r}, z)\exp\{-jkz\} = 0 \tag{2.4}$$

Consider the second term in equation (2.4). As $A(\mathbf{r}, z)\exp\{-jkz\}$ is a product of two functions in $z$, the second derivative with respect to $z$ gives four terms. Taking the first derivative with respect to $z$ gives

$$\frac{\partial}{\partial z}(A(\mathbf{r}, z)\exp\{-jkz\}) = A(\mathbf{r}, z)(-jk)\exp\{-jkz\} + \exp\{-jkz\}\frac{\partial A(\mathbf{r}, z)}{\partial z} \tag{2.5}$$

Taking the second derivative with respect to $z$, the derivative of equation (2.5),

$$\frac{\partial^2}{\partial z^2}(\mathbf{A}(\mathbf{r}, z)\exp\{-jkz\})$$

$$= \mathbf{A}(\mathbf{r}, z)(-k^2)\exp\{-jkz\} + (-jk)\exp\{-jkz\}\frac{\partial \mathbf{A}(\mathbf{r}, z)}{\partial z}$$

$$+ \exp\{-jkz\}\frac{\partial^2 \mathbf{A}(\mathbf{r}, z)}{\partial^2 z} + \frac{\partial \mathbf{A}(\mathbf{r}, z)}{\partial z}(-jk)\exp\{-jkz\} \qquad (2.6)$$

At right side of equation (2.6) we neglect the third term by using the slowly varying amplitude approximation $\partial^2 A/\partial z^2 \ll \partial A/\partial z$, equation (8.13) [176], combine the second and fourth term and cancel the first term on substituting into equation (2.4), to get the paraxial Helmholtz equation (with $\nabla_{\mathbf{r}}^2 = \partial^2/\partial x^2 + \partial^2/\partial y^2$)

$$\nabla_{\mathbf{r}}^2 \mathbf{A}(\mathbf{r}, z) - 2jk\frac{\partial \mathbf{A}(\mathbf{r}, z)}{\partial z} = 0 \qquad (2.7)$$

Equation (2.7) provides $\partial \mathbf{A}/\partial z$, which describes how the envelope $\mathbf{A}$ propagates in $z$. Integrating equation (2.7) gives a parabolic wave approximation (Section 1.1) to the spherical wave close to the axis:

$$\mathbf{A}(\mathbf{r}) = \frac{\mathbf{A}}{z}\exp\left\{-jk\frac{\rho^2}{2z}\right\} \qquad (2.8)$$

where $\rho^2 = x^2 + y^2$. That equation (2.8) is the solution to equation (2.7) can be verified by substitution and appropriate differentiation [148].

The Gaussian beam is obtained from the parabolic wave by transforming $z$ by a purely imaginary amount $jz_0$, where $z_0$ is called the Rayleigh range,

$$q(z) = z + jz_0 \qquad (2.9)$$

Therefore, we replace $z$ in equation (2.8) by $q(z)$ defined in equation (2.9):

$$A(\mathbf{r}) = \frac{A_1}{q(z)}\exp\left\{-jk\frac{\rho^2}{2q(z)}\right\} \qquad (2.10)$$

We now show that the complex term $1/q(z)$, arising twice in equation (2.10), fully describes the propagation of a Gaussian beam at $z$: its real part specifying the radius of curvature $R(z)$ of the phase front and the imaginary part specifying the spot size $W(z)$, which is the radius of the spot at $1/e$ of the peak. Separating the reciprocal of equation (2.9) into real and imaginary parts,

$$\frac{1}{q(z)} = \frac{1}{z + jz_0} = \frac{z - jz_0}{z^2 + z_0^2} = \frac{z}{z^2 + z_0^2} - j\frac{z_0}{z^2 + z_0^2} \qquad (2.11)$$

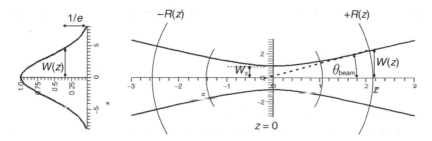

**FIGURE 2.1** Gaussian beam propagation.

The reciprocal of the real part, second to the last term in equation (2.11), defines the radius of curvature $R(z)$:

$$R(z) = z\left[1 + \left(\frac{z_0}{z}\right)^2\right] \tag{2.12}$$

This describes how the phase front associated with the radius of curvature $R(z)$ expands and contracts with distance (Figure 2.1). The imaginary part of the reciprocal of last term in equation (2.11) times $\lambda/\pi$ defines the beam radius $W(z)$ from

$$W^2(z) = \frac{\lambda}{\pi}z_0\left[1 + \left(\frac{z}{z_0}\right)^2\right] \tag{2.13}$$

Using equations (2.12) and (2.13), we can rewrite equation (2.11) in terms of $R$ and $W$:

$$\frac{1}{q(z)} = \frac{1}{R(z)} - j\frac{\lambda}{\pi}\frac{1}{W^2(z)} \tag{2.14}$$

Note that $\pi W^2$ is the area of the beam falling inside amplitude $1/e$ of its peak. From equation (2.13), for $z = 0$, the waist of the beam, the narrowest part, is

$$W_0^2 = \frac{\lambda}{\pi}z_0 \tag{2.15}$$

Substituting equation (2.15) into equation (2.13) gives the beam radius $W(z)$ in terms of its waist $W_0$ and distance $z$ from the waist (Figure 2.1),

$$W^2(z) = W_0^2\left[1 + \left(\frac{z}{z_0}\right)^2\right] \tag{2.16}$$

From equations (2.11) and (2.16), the amplitude and the phase of $1/q(z)$ can be written as

$$\frac{1}{\sqrt{z^2 + z_0^2}} = \frac{1}{z_0}\frac{W_0}{W}(z) \quad \text{and} \quad \zeta(z) = \tan^{-1}\frac{z}{z_0} \tag{2.17}$$

where we used equations (2.13) and (2.15) for the first equality.

Consider equation (2.10) for the envelope of the Gaussian beam. In the amplitude part of equation (2.10), replace $1/q(z)$ with real and imaginary parts of $1/q(z)$ from equation (2.17). In the phase part of equation (2.10), replace the real and imaginary parts of $1/q(z)$ from equation (2.14). The resulting expression for the complex envelope of a Gaussian beam can be written as

$$A(\mathbf{r}) = \frac{A_1}{z_0}\frac{W_0}{W(z)}\exp\{j\zeta(z)\}\exp\left[-jk\frac{\rho^2}{2}\left(\frac{1}{R(z)} - j\frac{\lambda}{\pi}\frac{1}{W^2(z)}\right)\right] \tag{2.18}$$

Hence, by using $k = 2\pi/\lambda$ (in air) and substituting the complex envelope, $A(\mathbf{r})$ in equation (2.18), into the complex amplitude of the Gaussian beam field, equation (2.1),

$$U(\mathbf{r}) = \frac{A_1}{z_0}\frac{W_0}{W(z)}\exp\left[-\frac{\rho^2}{W^2(z)}\right]\exp\left[-jkz - jk\frac{\rho^2}{2R(z)} + j\zeta(z)\right] \tag{2.19}$$

The properties of the Gaussian beam are apparent in Figure 2.1 and elaborated in detail elsewhere [148, 176].

## 2.1.2    Gaussian Beam with ABCD Law

Interestingly, knowledge of the $2 \times 2$ matrix for an optical element, labeled clockwise from top left element as $ABCD$ (Section 1.2.6), is sufficient to determine the change from $q_1$ to $q_2$ describing a Gaussian beam [148, 176] propagating through the optical element, equation (2.14). For propagation through an optical element, converting rays to Gaussian beams is accomplished by the $ABCD$ law

$$q_2 = \frac{q_1 A + B}{q_1 C + D} \tag{2.20}$$

For paraxial waves, equation (2.20) is proved analytically in Ref. [176] and by induction in Ref. [148]. In induction, each element in the ray path, no matter how complex, may be discretized along the $z$ axis into thin slivers that either shift the ray position, matrix (1.20), or change the ray slope (1.21). Now we show that equation (2.20) will give the correct change in Gaussian beam parameter $q$ for both these cases. So, by induction, for any optical element, equation (2.20) will provide the information on the propagation of a Gaussian beam.

For a Gaussian beam passing through a slab of constant refractive index, the slope of the ray is constant, but the location of the ray changes with distance (the ray can be considered to be along the $1/e$ amplitude edge of a Gaussian beam). From equation (2.9) with $q_1$ at the input and $q_2$ at the output,

$$q_2 = z_2 + jz_0 \quad \text{and} \quad q_1 = z_1 + jz_0 \qquad (2.21)$$

Subtracting one equation from the other in equation (2.21) for a thin sliver gives

$$q_2 - q_1 = z_2 - z_1 = d \qquad (2.22)$$

where $d$ is the thickness of the slab.

The matrix elements $A = 1$, $B = d$, $C = 0$, and $D = 1$ from equation (1.20) for propagating across a slab of width $d$ are inserted into the $ABCD$ law, equation (2.20), to give

$$q_2 = \frac{q_1 1 + d}{q_1 0 + 1} = q_1 + d \qquad (2.23)$$

which is the same as equation (2.22), proving the $ABCD$ law for propagation across a slab.

For a Gaussian beam passing through a dielectric interface with change in refractive index from $n_1$ to $n_2$, the beam radius $W$ is the same on both sides of the interface, but the ray slope changes at the interface due to refraction. From equation (2.14),

$$\frac{1}{q_2} = \frac{1}{R_2} - j\frac{\lambda_2}{\pi}\frac{1}{W^2} \quad \text{and} \quad \frac{1}{q_1} = \frac{1}{R_1} - j\frac{\lambda_1}{\pi}\frac{1}{W^2} \qquad (2.24)$$

Consider a ray along the edge of the Gaussian beam ($1/e$ amplitude) (Figure 2.2) incident on the interface at angle $\theta_1$ with the axis and having radius of curvature $R_1$ and spot radius $W$ at the interface. After refraction at the interface, while the beam radius $W$ is unchanged, the angle with the horizontal becomes $\theta_2$ and the radius of

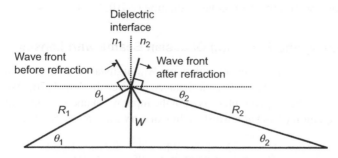

**FIGURE 2.2**   Refraction of Gaussian beam at dielectric interface.

curvature becomes $R_2$. Then

$$\frac{W}{R_1} = \sin \theta_1 \quad \text{and} \quad \frac{W}{R_2} = \sin \theta_2. \tag{2.25}$$

Dividing one equation by the other and using Snell's law gives

$$\frac{R_2}{R_1} = \frac{\sin \theta_1}{\sin \theta_2} = \frac{n_2}{n_1} \quad \text{or} \quad \frac{1}{R_2} = \frac{n_1}{n_2}\frac{1}{R_1} \tag{2.26}$$

As wavelength is inversely proportional to refractive index ($\lambda = c/(n(\text{freq}))$), we can write

$$\lambda_2 = \frac{n_1}{n_2}\lambda_1 \tag{2.27}$$

Substituting $R_2$ from equation (2.26) and $\lambda_2$ from equation (2.27) into the first equation in equation (2.24) shows, in the first equation, the change in Gaussian beam parameter to $q_2$,

$$\frac{1}{q_2} = \frac{n_1}{n_2}\left(\frac{1}{R_1} - j\frac{\lambda_1}{\pi}\frac{1}{W^2}\right) \quad \text{or} \quad \frac{1}{q_2} = \frac{n_1}{n_2}\frac{1}{q_1} \tag{2.28}$$

The second equation in equation (2.24) is used to obtain the second equation in equation (2.28).

Inserting the matrix elements from equation (1.21) $A = 1$, $B = 0$, $C = 0$, $D = n_1/n_2$ and using the $ABCD$ law, equation (2.20),

$$q_2 = \frac{q_1 1 + 0}{q_1 0 + n_1/n_2} = q_1\frac{n_2}{n_1} \quad \text{or} \quad \frac{1}{q_2} = \frac{n_1}{n_2}\frac{1}{q_1} \tag{2.29}$$

which is the same as equation (2.28), proving the $ABCD$ law for crossing from one refractive index to another.

Thus, by induction, as any optical element may be written in terms of thin slivers that are either a constant refractive index or a dielectric interface between two different media, we have proven the $ABCD$ law, equation (2.20), for all cases.

### 2.1.3 Forming and Receiving Gaussian Beams with Lenses

Figure 2.3 shows a Gaussian beam with waist $W_{01}$ at distance $d_1$ in front of a convex lens that focuses light to waist $W_{02}$ at distance $d_2$ after the lens [176]. We label the complex Gaussian beam parameters as $q_1$ at the input waist and $q_2$ at the output waist. Then, from equation (2.14) at a waist, radius of curvature $R_1 = \infty$, $1/R_1 \to 0$,

$$\frac{1}{q_1} = -j\frac{\lambda}{\pi}\frac{1}{W_{01}^2} = \frac{1}{jz_1} \quad \text{or} \quad q_1 = jz_1 \tag{2.30}$$

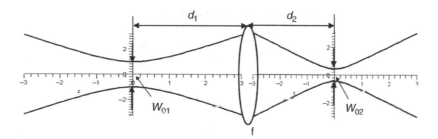

**FIGURE 2.3**   Focusing a Gaussian beam.

where we used the confocal beam parameter $z_1$ that describes the distance at which the waist expands by $\sqrt{2}$, that is, $z = z_0$ in equation (2.16). Similarly, for $R_2 = \infty$, $1/R_2 \to 0$,

$$\frac{1}{q_2} = -j\frac{\lambda}{\pi}\frac{1}{W_{02}^2} = \frac{1}{jz_2} \quad \text{or} \quad q_2 = jz_2 \tag{2.31}$$

For a medium of refractive index $n$, we can replace $\lambda$ with $\lambda/n$.

For a ray passing from input waist to output waist, in order to define $A$, $B$, $C$, and $D$, we write the product in reverse order (the order of occurrence) of the three propagation matrices: to the lens through distance $d_1$, equation (1.20); through the lens, equation (1.22); and to the output waist through distance $d_2$,

$$\begin{bmatrix} A & B \\ C & D \end{bmatrix} = \begin{bmatrix} 1 & d_2 \\ 0 & 1 \end{bmatrix} \begin{bmatrix} 1 & 0 \\ \frac{1}{-f} & 1 \end{bmatrix} \begin{bmatrix} 1 & d_1 \\ 0 & 1 \end{bmatrix} = \begin{bmatrix} 1 - \frac{d_2}{f} & d_1 + d_2 - \frac{d_1 d_2}{f} \\ -\frac{1}{f} & 1 - \frac{d_1}{f} \end{bmatrix} \tag{2.32}$$

We use $A$, $B$, $C$, and $D$ from equation (2.32) with equations (2.30) and (2.31) to insert into the $ABCD$ law, equation (2.20). We include the unimodal property of the matrices, $AD - BC = 1$. The Gaussian beam complex propagation parameter $q$ is updated from $q_1$ to $q_2$ during propagation from the input to the output waists, as shown in Figure 2.3. $q_1$ and $q_2$ are represented by the respective confocal parameters $z_1$ and $z_2$ from equations (2.30) and (2.31),

$$jz_2 = \frac{jz_1 A + B}{jz_1 C + D} = \frac{(jz_1 A + B)(-jz_1 C + D)}{(z_1^2 C^2 + D^2)} = \frac{ACz_1^2 + BD + jz_1}{C^2 z_1^2 + D^2} \tag{2.33}$$

From equations (2.30) and (2.31), the confocal parameters $z_1$ and $z_2$ in terms of input and output spot sizes $W_{01}$ and $W_{02}$ in air ($n = 1$) are, respectively,

$$z_1 = \frac{\pi W_{01}^2}{\lambda} \quad \text{and} \quad z_2 = \frac{\pi W_{02}^2}{\lambda} \tag{2.34}$$

Equating real parts of either side of equation (2.33) gives

$$ACz_1^2 + BD = 0 \quad \text{or} \quad z_1^2 = -\frac{BD}{AC} \tag{2.35}$$

and equating imaginary parts of either side of equation (2.33) gives

$$z_2 = \frac{z_1}{C^2 z_1^2 + D^2} \tag{2.36}$$

To find the relation between the output spot size $W_{02}$ and the input spot size $W_{01}$, eliminating $z_1^2$ from equations (2.35) and (2.36), using the unimodal property of the matrices, $AD - BC = 1$, and assigning $A = 1 - (d_2/f)$ and $D = 1 - (d_1/f)$ from equation (2.32) gives

$$z_2 = \frac{A}{D}z_1 = \frac{d_2 - f}{d_1 - f}z_1 \quad \text{or from equation (2.34)} \quad W_{02}^2 = \frac{d_2 - f}{d_1 - f}W_{01}^2 \tag{2.37}$$

We would like the location and the spot size at the output as a function of only input parameters, hence we need $d_2$ as function of $d_1$ from equation (2.37). From equation (2.35), inserting $ABCD$ elements from equation (2.20),

$$z_1^2 = -\frac{BD}{AC} = -\frac{(d_1 + d_2 - (d_1 d_2)/f)(1 - d_1/f)}{(1 - d_2/f)(-1/f)}$$

$$= (d_1 f + d_2 f - d_1 d_2)\left(\frac{d_1 - f}{d_2 - f}\right) \tag{2.38}$$

$$(d_2 - f)z_1^2 = -(d_1 - f)^2(d_2 - f) + f^2(d_1 - f) \tag{2.39}$$

Hence, the location $d_2$ of the output waist is given by

$$(d_2 - f) = \left(\frac{f^2}{z_1^2 + (d_1 - f)^2}\right)(d_1 - f) \tag{2.40}$$

Substituting $(d_2 - f)/(d_1 - f) = W_{02}^2/W_{01}^2$ from equation (2.37) into equation (2.40), the waist radius $W_{02}$ at the output is

$$W_{02}^2 = \left(\frac{f^2}{z_1^2 + (d_1 - f)^2}\right)W_{01}^2 \tag{2.41}$$

Some special cases of Figure 2.3 are summarized [176].

(a) A point source has input spot size $W_{01} = 0$ or $z_1 = 0$ from equation (2.34). Then equation (2.40) with $z_1 = 0$ gives $(d_2 - f)(d_1 - f) = f^2$ or the lens law $1/d_1 + 1/d_2 = 1/f$.

(b) For a plane wave, $z_1 = \infty$, from equation (2.40), $d_2 = f$. The plane wave is focused to a point in the back focal plane.

(c) For the input waist at the front focal plane $d_1 = f$, from equation (2.40), the output waist is at the back focal plane.

(d) In a laser weapon (Section 12.2), the approximate plane wave at the source corresponding to an infinite radius of curvature, $d_1 = 0$, feeds directly into a lens or mirror for which the focal length can be selected to focus the light onto a waist on the target at distance $d_2$. Focusing onto a waist of the Gaussian beam can provide maximum intensity at the target spot. Similarly, in a laser warning device (Section 14.2.2), mounted on most military vehicles, the lens focuses a parallel threatening laser beam to a detector array at a distance $d_2$ for detection and direction estimation. For both cases, equation (2.40) with $d_1 = 0$ gives the distance between lens and target or detector array $d_2$:

$$(d_2 - f) = \frac{f^2}{z_1^2 + f^2}(-f) \quad \text{or} \quad d_2 = \frac{f}{1 + (f/z_1)^2} \tag{2.42}$$

where the confocal distance $z_1$ may be obtained by using $W_{01}$ and $\lambda$ in equation (2.34). From equation (2.41), the beam size at the array for $d_1 = 0$ is

$$W_{02} = \frac{f}{\sqrt{z_1^2 + f^2}} W_{01} = \frac{f/z_1}{\sqrt{1 + (f/z_1)^2}} W_{01} \tag{2.43}$$

## 2.2 POLARIZATION

Polarization is concerned with the direction of motion of the electric field vector in the transverse plane for electromagnetic wave propagation. The polarization angle $\psi$ is in a direction transverse to the direction of propagation for electromagnetic waves and is measured with reference to the $x$ axis. Its sign is defined while looking in the direction of propagation. So for a conventional western right-hand coordinate system (for which the right-hand rule may be used), we draw the transverse $x$ axis as vertical and the transverse $y$ axis as horizontal when looking in the $z$ propagation direction. Any polarization may be written as the sum of two orthogonal direction polarizations; for example, we write the waves for the orthogonal polarizations as [16, 83]

$$E'_x = E_x \cos(\tau + \phi_x) \quad \text{and} \quad E'_y = E_y \cos(\tau + \phi_y) \tag{2.44}$$

where $\tau = \omega t - k_z z$, and $\phi_x$ and $\phi_y$ are the phase lags for $E'_x$ and $E'_y$, respectively. The phase $\phi$ is a function of time and should not be confused with $\psi$, the angle of polarization. By expanding the angle sums in equation (2.44),

$$\frac{E'_x}{E_x} = \cos(\tau)\cos(\phi_x) - \sin(\tau)\sin(\phi_x), \qquad \frac{E'_y}{E_y} = \cos(\tau)\cos(\phi_y) - \sin(\tau)\sin(\phi_y)$$

(2.45)

Multiplying the first equation (2.45) by $\sin(\phi_y)$ and the second equation by $\sin(\phi_x)$ allows the last terms in the resulting equations to cancel on subtraction,

$$\frac{E'_x}{E_x}\sin(\phi_y) - \frac{E'_y}{E_y}\sin(\phi_x) = \cos(\tau)\sin(\Delta\phi)$$

$$\frac{E'_x}{E_x}\cos(\phi_y) - \frac{E'_y}{E_y}\cos(\phi_x) = \sin(\tau)\sin(\Delta\phi) \qquad (2.46)$$

where $\Delta\phi = \phi_y - \phi_x$. Squaring and adding gives

$$\left(\frac{E'_x}{E_x}\right)^2 + \left(\frac{E'_y}{E_y}\right)^2 - 2\frac{E'_x}{E_x}\frac{E'_y}{E_y}\cos(\Delta\phi) = \sin^2(\Delta\phi) \qquad (2.47)$$

This is an ellipse since the associated determinant is nonnegative; a typical polarization ellipse is shown in Figure 2.4a. If the ratio of $E_x$ to $E_y$ changes, the polarization angle $\psi$ changes. When the minor axis goes to zero, this reverts to linear polarization (Figure 2.4b), and when $E_x = E_y$, the ellipse reverts to a circle for circular polarization.

For *linear polarization*, the angle $\psi$, measured from direction $x$, is such that the two orthogonal polarizations in the transverse plane are $\Delta\phi = \phi_y - \phi_x = 0$ for linear polarizations from $0 \le \psi < \pi/2$ and $\Delta\phi = \pi$ for linear polarizations from $\pi/2 \le \psi < \pi$. Note that because $E$ field oscillations range from one side of the origin to the other, polarization repeats after rotating by $\psi = \pi$, while, in contrast, regular angles such as phase $\phi$ repeat after $2\pi$.

For *circular polarization*, $\Delta\phi = \phi_x - \phi_y = \pm\pi/2$. When $x$ direction peaks just before $y$ direction peaks, the $y$ direction phase lags that of the $x$ direction by $+\pi/2$,

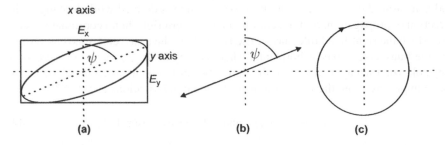

(a)                     (b)                     (c)

**FIGURE 2.4** Polarization: (a) general polarization ellipse, (b) special linear case, and (c) special circular case.

resulting in right circular (clockwise) polarization (Figure 2.4a). When the $x$ direction peaks just after the $y$ direction peaks, the $y$ direction phase leads that of the $x$ direction, or lags by $-\pi/2$, resulting in left circular (anticlockwise) polarization.

## 2.2.1  Wave Plates or Phase Retarders

Polarization may be changed with a slice of an anisotropic crystal called a wave plate or a phase retarder [83]. Typically, we use uniaxial crystals for which one of the three orthogonal axes has a different refractive index from the other two and this direction is lined up with the flat surface of the crystal slice. The single unique axis is referred to as extraordinary (subscript e), while the other two are referred to as ordinary (subscript o). The extraordinary direction is known as the optic axis and is marked with a dot labeled $OA$ on the edge of a wave plate. The velocity of propagation of light $v$ depends on the refractive index $v = c/n$, where $c$ is the speed of light in a vacuum. Therefore, waves with polarization ($E$ field oscillation) in the direction of the optic axis will see a refractive index $n_e$, while waves with polarization at right angles to the optic axis will see refractive index of $n_o$. For a *fast* crystal, the light polarized in the direction of the optic axis will travel faster than that polarized in orthogonal directions; this means $n_e < n_o$. Consequently, at the crystal output, the phase of the extraordinary light will be ahead of that for an orthogonal direction. For a slow crystal, $n_o < n_e$, the light with optic axis direction polarization will travel more slowly than that with the orthogonal polarizations and have a phase behind that of the orthogonal polarizations.

### 2.2.1.1  *Half-Wave Plate*  Consider a uniaxial crystal of depth $d$ and optical axis $OA$ on the surface in direction $x$ as shown in Figure 2.5a. The refractive indices are $n_x \equiv n_e$, and $n_y = n_z \equiv n_o$.

When light travels through a material of width $d$ and refractive index $n$, the path length, equivalent distance traveled through air, is $nd$. Consider that a linearly polarized wave strikes the crystal at normal incidence to the crystal and with a polarization

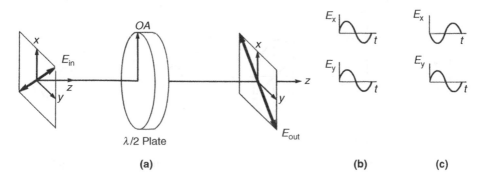

**FIGURE 2.5**   Half-wave plate: (a) operation, (b) input wave components, and (c) output wave components.

at $\psi = 45° \equiv \pi/4$ with respect to the optic axis. For a half-wave plate, the width $d$ of the crystal is such that the extraordinary ray is delayed or advanced by half a wavelength with respect to the ordinary wave, $|n_o - n_e|d = \lambda/2$. Then, there will be a phase difference between the orthogonal polarizations in $x$ and $y$ of $\Delta\phi = \pi$ radians at the crystal output, which leads to a change in polarization of $\psi = \pi/2$. As $d$ may be too thin for safe handling and because of the cyclic nature of waves, the length can be increased by any integer multiple $m$ of wavelengths,

$$|n_o - n_e|d = \lambda/2 + m\lambda \qquad (2.48)$$

If the overall width of the plate $d$ is to be greater than $d'$, the minimum width for handling, $d \geq d'$, we first solve for next highest integer $m$ from $m =$ ceiling ($|n_o - n_e|d'/\lambda$) (see maple ceil() function) and then use equation (2.48) with this integer value of $m$ to find the correct width $d$. Figure 2.5b shows the components of the input wave along $x$ and $y$. The input waves are synchronized so that they both increase together to produce the vector $E_{in}$. The output waves are exactly $\pi$ out of phase, so that when one increases, the other decreases as in Figure 2.5c. Together they form the vector $E_{out}$, which represents polarization $\pi/2$ different from $E_{in}$. A linear polarizer after the crystal is set to pass or block light polarized at $\pi/4$. A voltage into an electro-optic crystal provides a mechanism for switching the output on and off for displays.

### 2.2.1.2 *Quarter-Wave Plate*

If the depth $d$ of the crystal is such that the extraordinary ray is delayed or advanced by $|n_o - n_e|d = \lambda/4$, relative to the ordinary ray, there will be a phase difference between them of $\Delta\phi = \pi/2$ and the crystal is known as a quarter-wave plate as shown in Figure 2.6. The quarter-wave plate causes a $\Delta\phi = \pi/2$ phase shift of the phase $\phi$ of the $x$ polarization relative to the $y$ polarization. If a plane monochromatic wave strikes the crystal in a normal incidence,

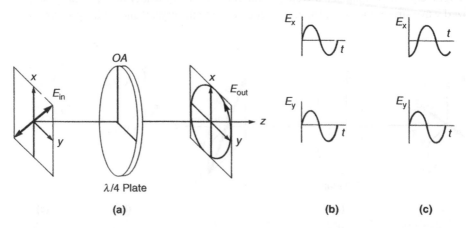

**FIGURE 2.6** Quarter-wave plate: (a) operation, (b) Input wave components, and (c) output wave components.

and it has a polarization at $\psi = 45° \equiv \pi/4$ with respect to the optic axis, then there will be equal components as extraordinary and ordinary rays. If the crystal has a fast axis such as calcite, the extraordinary wave will travel faster than the ordinary wave. The $\Delta\phi = \pi/2$ change in phase will cause a linearly polarized beam to be converted into a circularly polarized one, as shown in Figure 2.6b and at $E_{out}$ (Figure 2.6a). The input components are shown in Figure 2.6b and the output components with the $x$ component delayed by 90° relative to the $y$ component in Figure 2.6c. With the $x$ component delayed relative to the $y$ component, the rotation is left circular (anticlockwise). In general, the direction of rotation of the circularly polarized beam depends on the direction of the linear polarization relative to the optical axis and on whether the crystal is fast or slow.

Note that two identical quarter-wave plates in sequence add to become a halfwave plate. For example, this is used to separate incoming and outgoing light beams in compact disk players. In the latter, polarized light from a laser diode passes through a polarizing beam splitter and a quarter-wave plate before reflecting from a disk to distinguish the presence of raised spots ("zero") that scatter light and the absence of raised spots that reflect light. The reflection for a "one" (flat surface) has a changed polarization because the direction $z$ is reversed to $-z$. On passing back through the quarter-wave plate, the second passage through the plate makes the combination of two plates look like a single half-wave plate. So polarization is now at 90° to the incoming light and will be deflected at the polarizing beam splitter to a photodetector and does not return to the laser diode where it could interfere with laser oscillation (Section 6.2). This structure allows a common path for input and output beams, which alleviates alignment and avoids directional isolators. Directional isolators block light from traveling along specific paths by means of Faraday rotators [176] and are used in the 94 GHz radar quasi-optical duplexer (Section 16.2.4) because of the frequency and power involved.

In summary, this is also true if a single quarter-wave plate is used with a mirror so that the beam travels back through the same plate rather than through a second plate. In other words, two adjoining quarter-wave plates will form a half-wave plate. As mentioned this is used in a compact disk player in conjunction with a polarizing beam splitter, which reflects orthogonal polarizations in different directions, to separate the source light emitted from the laser diode from light reflected from the disk.

## 2.2.2 Stokes Parameters

There are several ways to represent polarization: Stokes parameters, Jones calculus, Mueller matrices, and Poincaré sphere. Stokes parameters are obtained by direct measurement with a polarizer and a quarter-wave plate at a transverse plane in space ($\exp\{-jkz\}$ is ignored). We represent the two fields in the orthogonal directions to the direction of propagation as

$$E'_x = \text{Re}[E_x \exp\{j(\omega t + \phi_x(t))\}]$$
$$E'_y = \text{Re}[E_y \exp\{j(\omega t + \phi_y(t))\}] \tag{2.49}$$

The four Stokes equations are (with $\Delta\phi = \phi_y - \phi_x$)

$$s_0 = |E_x|^2 + |E_y|^2$$
$$s_1 = |E_x|^2 - |E_y|^2$$
$$s_2 = 2|E_x||E_y|\cos(\Delta\phi)$$
$$s_3 = 2|E_x||E_y|\sin(\Delta\phi) \tag{2.50}$$

These are measured as follows [16]. If $I(\theta)$ uses a linear polarizer at an angle $\theta$ to measure intensity at angle $\theta$, and $Q(\theta)$ uses a quarter-wave plate followed by a linear polarizer at angle $\theta$, then the Stokes parameters are measured by

$$s_0 = I(0) + I(\pi/2)$$
$$s_1 = I(0) - I(\pi/2)$$
$$s_2 = I(\pi/4) - I(3\pi/4)$$
$$s_3 = Q(\pi/4) - Q(3\pi/4) \tag{2.51}$$

In the Stokes parameters, $s_0$ represents the total power where $s_0^2 = s_1^2 + s_2^2 + s_3^2$. So only three of the four Stokes parameters are independent and need be measured. $s_1$ represents excess power in $x$ direction polarization over $y$ direction polarization, $s_2$ represents the amount of $\pi/4$ polarized power over that in $-\pi/4$, and $s_3$ represents the amount of right circular polarized power over left circular polarized power.

### 2.2.3 Poincaré Sphere

Polarimeters that measure polarization often display the polarization on a sphere [16]. The sphere, introduced by Poincaré, provides a convenient way of representing all possible polarizations. It is interesting that for analysis to obtain simpler results and for advanced research, Poincaré used spherical geometry [16], in which, for example, the three angles of a triangle (placed on a sphere) no longer add to 180°. Every possible polarization is represented by one point on the Poincaré sphere and every point on the sphere represents one polarization (a one-to-one mapping) (Figure 2.7a) [29]. Points on sphere are marked with $x, y, z$ coordinates. Polarimeters may show the whole sphere or only one quadrant with a note to say which quadrant is displayed. A Siemens VPI simulator, popular for optical network design, allows the sphere to be turned with a mouse to view any quadrant.

Points $P$ on the sphere along the equator represent linear polarizations at different angles $\psi$, at left linear horizontal polarization (LHP), and opposite at right on the sphere linear vertical polarization (LVP). Note that opposite points on the sphere are uncorrelated. Twice the angle of the polarization, or $2\psi$, is the azimuth of the point $P$ (Figure 2.7b). As $P$ moves up from the equator to elevation $2\chi$ (Figure 2.7a and b), the minor axis of the right circular ellipse increases until the polarization becomes right circular at the north pole. If we return back to the equator, on passing through the equator, the minor axis inverts to produce left circular elliptical

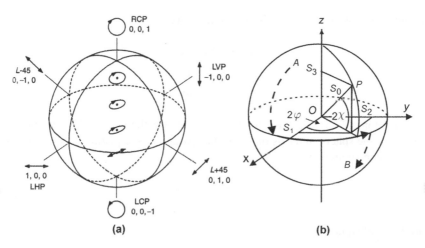

**FIGURE 2.7**   Poincaré sphere: (a) polarizations and (b) Stokes parameters on Poincaré sphere and polarization control.

polarization. The angle $\gamma$ subtended at the center of the sphere between two different polarizations $P_1$ and $P_2$ gives the correlation between the two polarizations:

$$\rho = \cos\left(\frac{\gamma}{2}\right) \tag{2.52}$$

For example, if points $P_1$ and $P_2$ are coincident, $\gamma = 0$ and correlation is $\rho = \cos(0) = 1$ or perfect correlation. As mentioned, if points $P_1$ and $P_2$ are on opposite sides of the sphere, $\gamma = \pi$ and correlation is $\rho = \cos(\pi/2) = 0$ for uncorrelated points. Interference patterns improve as the correlation between interfering beams increases. If light loses polarization, it will move down the line toward the center of the sphere. The origin of the sphere has no distinct polarization (it has all polarizations equally). If the polarization drifts with time, it will create a time trajectory on the surface of the sphere. If many frequencies are present, the polarization will differ and there will be a point for each, so for a pulse with a continuum of frequencies, a trace will appear. Due to dispersion, the trace will change shape with propagation.

### 2.2.4   Finding Point on Poincaré Sphere and Elliptical Polarization from Stokes Parameters

A shown in Figure 2.7b, we plot the Stokes parameters $s_1$, $s_2$, and $s_3$ on the $x$, $y$, and $z$ axes, respectively, where these axes are drawn over the sphere. A polarization point $P$ on the sphere can now be described with rectangular coordinate $(s_1, s_2, s_3)$ or spherical coordinate $(s_0, 2\chi, 2\psi)$, where the radius of the sphere is the total power $s_0$, $2\chi$ is the elevation, and $2\psi$ is the azimuthal angle. Note that the spherical coordinate system is that used for the earth; electrical engineers typically use angle from north pole rather than elevation. By propagating the vector from origin to $P$ down to the

$x$, $y$, and $z$ axes, we obtain the following equations:

$$s_1 = s_0 \cos 2\chi \cos 2\psi$$
$$s_2 = s_0 \cos 2\chi \sin 2\psi$$
$$s_3 = s_0 \sin 2\chi \tag{2.53}$$

From equations (2.53), we can derive expressions for the spherical coordinates for the point on the sphere in terms of the Stokes parameters, equation (2.50). From the first and second equations,

$$\tan 2\psi = \frac{s_2}{s_1} \tag{2.54}$$

From the third equation,

$$\sin 2\chi = \frac{s_3}{s_0} \tag{2.55}$$

Hence, a point in spherical equations on the Poincaré sphere can be equated to the one in rectangular coordinates from the Stokes parameters and vice versa.

We also know the difference in phase angle $\Delta\phi$ between the $x$ and $y$ wave components from the Stokes parameters, equation (2.50),

$$\tan(\Delta\phi) = \frac{s_3}{s_2} \tag{2.56}$$

We can also draw the elliptical polarization plot like that in Figure 2.4a by computing the circumscribing rectangle of size $2E_x$ by $2E_y$. From the first two Stokes equations (2.50),

$$|E_x|^2 = \left(\frac{s_0 + s_1}{2}\right)^{1/2}$$
$$|E_y|^2 = \left(\frac{s_0 - s_1}{2}\right)^{1/2} \tag{2.57}$$

The polarization angle $\psi$ was computed in equation (2.54).

## 2.2.5   Controlling Polarization

A polarization controller is commonplace in optical fiber telecommunications. Fiber is birefringent because it is not perfectly circularly symmetric. Hence, fiber has different refractive indices in two orthogonal cross-section directions; that is, vertical polarized waves will travel at a different speed from horizontal polarized waves, or indeed any arbitrary polarized wave will travel at a different speed from its opposite polarization. Worse still, any movement in the fiber will change the polarization.

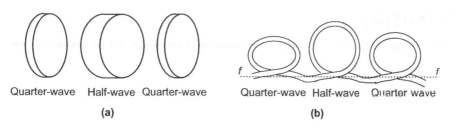

Quarter-wave  Half-wave  Quarter-wave          Quarter-wave  Half-wave  Quarter wave

(a)                                            (b)

**FIGURE 2.8** Polarization controller: (a) in free space with wave plates and (b) in optical fiber.

Consequently, unless more expensive polarization maintaining fiber is used, the polarization at any point in a fiber is generally unknown. Many telecommunication components are polarization sensitive, often because they are constructed from integrated optic components that typically have waveguides for which TE and TM modes propagate differently. The different speeds for orthogonal components cause a change in polarization similar to that for wave plates (Section 2.2.1). Hence, the component may have several dB different behavior for two opposite polarizations.

One solution to connecting a polarization-sensitive device to a fiber network is to insert a polarization scrambler in front of the device. Such a device spreads the polarization, moving a spot on the Poincaré sphere closer to the center or, equivalently, spreading it equally over the whole sphere—not an easy or inexpensive task. Another solution is to purchase a more expensive version of the component that is polarization insensitive. Optic devices can be made polarization insensitive by dividing the incoming wave into two orthogonal polarizations, treating each separately, and then combining the results, basically paying for two components. An attractive alternative is to use a polarization controller.

Polarization controllers come in many forms [131]. In free space, three or four wave plates can be used in sequence, as shown in Figure 2.8a. Figure 2.8a shows a sequence of quarter-wave, half-wave, and quarter-wave plates; the half-wave plate could be two quarter-wave plates (Section 2.2.1). Plates are arranged with their optic axis in the same direction and each plate is then rotated about its axis to arrange for a starting polarization $A$ in Figure 2.7b to be converted to a final polarization $B$. The first plate removes any circularity of polarization at point A, bringing the polarization to linear that falls on the equator of the sphere (Figure 2.7b). The half-wave plate moves the polarization around the equator to take on a different polarization angle but remain linear. The slope (polarization angle $\psi$) of the linear polarization is selected to conform with the major axis of the final polarization $B$. The last quarter-wave plate moves the polarization along a great circle toward a pole, up for right-hand polarization and down for left-hand polarization until the final polarization at point $B$ is reached.

Figure 2.8b shows how the same task is accomplished with fiber. A fiber loop when twisted about the main fiber axis puts stress on the fiber that changes the relative refractive indices for orthogonal directions. The loop is designed to create an action identical to quarter- and half-wave plates in Figure 2.8a. In Ref. [100], we have used this idea to form a very sensitive sensor.

# CHAPTER 3

# OPTICAL DIFFRACTION

Light from a laser, unlike that from incandescent sources, is close to a single frequency and can form a narrow beam because it is both temporally and spatially coherent. A laser pointer is a familiar example. By increasing the continuous or pulsed power from milliwatts to as high as megawatts or gigawatts, many military applications become feasible, for example, communications, ranging, target designation, guiding ordnance, dazzle, and destruction. Diffraction is the basic spatial coherence phenomenon that allows us to determine how rapidly a coherent beam spreads with distance, how fast a pulse spreads in time, and how sharply the beam can be focused, all critical in military systems. In Section 3.1, we describe the concept of diffraction and review 1D and 2D temporal and spatial Fourier transforms. In Section 3.2, we prove the uncertainty principle for Fourier transforms in time and in space and show how the latter may be used for simple spatial approximations for diffraction. In Section 3.3, we provide the formal equation derivation for scalar Fresnel and Fraunhofer diffraction; the scalar theory is a good approximation for 3D diffraction. In Section 3.4, we develop the theory of diffraction-limited imaging for determining how an aperture limits the quality of an image.

## 3.1  INTRODUCTION TO DIFFRACTION

We explain diffraction and review 1D and 2D Fourier transforms used for diffraction computations.

*Military Laser Technology for Defense: Technology for Revolutionizing 21st Century Warfare*,
First Edition. By Alastair D. McAulay.
© 2011 John Wiley & Sons, Inc. Published 2011 by John Wiley & Sons, Inc.

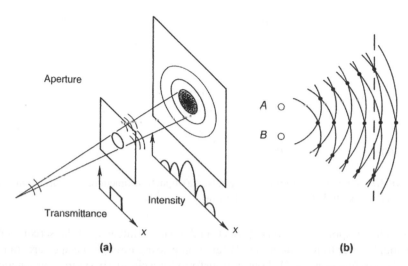

**FIGURE 3.1** Illustration of diffraction as per Huygens: (a) for a circular aperture and (b) two waves Interfering.

### 3.1.1 Description of Diffraction

Diffraction has been defined as behavior that cannot be explained by straight line rays [83]. Figure 3.1a shows such a case. If a coherent or single-frequency point source illuminates a card having a circular hole or aperture, a sharp shadow will not appear on a screen behind the card unless the screen is very close to the card. Instead, light passing through the aperture acts as a multitude of point sources, including those at the circular aperture edge. As a result, energy falls outside the shadow-casting region of the screen. The waves from these point sources interfere with each other as shown in Figure 3.1b, where points $A$ and $B$ may be considered to be point sources on opposite edges of the aperture. The spherical curves represent the peaks of the emanating waves at one instant of time. Constructive interference occurs where the peaks from the two sources coincide. Destructive interference occurs where a peak of one corresponds to a negative peak or dip of the other. The interference effect known as diffraction shows up as a series of diffraction rings surrounding the directly illuminated region on the screen (Figure 3.1a). The rings get fainter with size. In 1818, Fresnel won a prestigious French prize for showing the wave nature of light, thought to settle the controversy of whether light was basically particles, as Newton had conjectured, or waves as evidenced by interference. Since quantum mechanics in the early 20th century, it is known that light can be viewed as either a particle or a wave. At lower frequencies wave properties dominate and at higher frequencies particle properties dominate. Light is at the transition so that for understanding interference we assume waves and for lasers we need to assume particles.

   Figure 3.1a also shows the transverse light intensities across the center of the screen in the $x$ direction. At sufficient distance between aperture and screen, consistent with

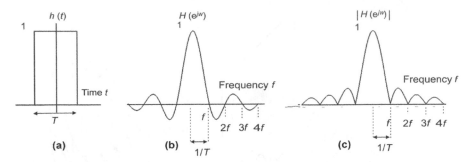

**FIGURE 3.2**   Fourier Transform of square pulse: (a) input time function, (b) Fourier transform of (a) and (c) magnitude Fourier transform of (a).

the Fraunhofer approximation [49] (Section 3.3.5), the intensity at the screen is the 2D spatial Fourier transform of the 2D function passing through the aperture, in this case, a Bessel function, the 2D Fourier transform for a circularly symmetric function. If a piece of film with an image is placed on the aperture, the 2D Fourier transform of the image would be obtained at the screen. The distance required is several meters. However, a converging source or the use of a lens permits the Fourier transform at much shorter distances [83].

### 3.1.2   Review of Fourier Transforms

First, we discuss temporal Fourier transforms in time, useful for pulses, and then the spatial Fourier transform used for diffraction. The one-dimensional Fourier transform transforms a function $f(t)$ of time $t$ in seconds into a function $F(\omega)$ of temporal frequency $\omega$ in radians per second:

$$\overbrace{F(w)}^{\text{output}} = \int_t \overbrace{f(t)}^{\text{input}} e^{-jwt} \, dt \tag{3.1}$$

Equation (3.1) may be viewed as correlating $f(t)$ against different frequencies to determine how much of each frequency $F(\omega)$ exists in the function $f(t)$. Temporal frequency $v = \omega/(2\pi)$ is measured in cycles per second or hertz. If a pulse of height unity (Figure 3.2a) has duration $T$, the transform from equation (3.1) is (Figure 3.2b)

$$F(w) = \int_{-T/2}^{T/2} \exp\{-j\omega t\} dt = \left. \frac{\exp\{-j\omega t\}}{-j\omega} \right|_{-T/2}^{T/2} = \frac{2\sin(\omega T/2)}{\omega} \tag{3.2}$$

The amplitude spectrum in equation (3.2) is zero or null when $\omega T/2 = n\pi$, where $n$ is an integer. The first null (the Rayleigh distance) occurs when $n = 1$, at $\omega T/2 = \pi$ or $\omega = 2\pi/T$, or in frequency at $f = \omega/(2\pi) = 1/T$, as shown in Figure 3.2b. The magnitude of the Fourier transform of a square pulse or boxcar is shown in Figure 3.2c. Intensity is magnitude squared.

A 1D electromagnetic wave $\cos(\omega t - k_x x)$ oscillates in time with angular frequency $\omega$ radians per second and oscillates in 1D space $x$ with spatial angular frequency $k_x$ radians per meter. In an analogous transverse water wave, a cork floating on a wave bobs up and down with frequency $\omega$ radians per second and a flash photograph shows the spatial wave oscillating with $k_x$ radians per meter in $x$ direction. So the 1D Fourier transform in the spatial domain may be written by analogy with equation (3.1) by replacing time $t$ by space variable $x$ and angular frequency in time $\omega$ radians per second by angular frequency in space $k_x$ radians per meter:

$$F(k_x) = \int_x f(x) e^{-jk_x x}\, dx \qquad (3.3)$$

where $k_x$ is the wave number in the $x$ direction in radians per meter and $x$ is distance in meters. Spatial frequency $f_x = 2\pi k_x$ has units of cycles per meter (or lines per meter). Physical space has more than one dimension; consequently, higher dimension transforms are of interest.

In considering beams or propagation of images between planes, we will consider 2D spatial Fourier transforms. A 1D spatial Fourier transform, equation (3.3), may be extended to 2D as follows:

$$\overbrace{F(k_x, k_y)}^{\text{2D output}} = \int_x \int_y \overbrace{f(x, y)}^{\text{2D input}}\, \overbrace{e^{-j(k_x x + k_y y)}}^{\text{plane wave}}\, dx dy \qquad (3.4)$$

where spatial frequencies $k_x$ and $k_y$ are in radians per meter in $x$ and $y$ directions, respectively. Note that the input wave pattern is decomposed into a set of plane waves with different propagation directions $(k_x, k_y)$ and amplitudes. The 2D input pattern is correlated with each plane wave to give $F(k_x, k_y)$, the amount of each plane wave present in the input. A square shaped beam or source of width and height $S$ will have a transform similar to that shown in Figure 3.2 in $x$ and $y$ in the 2D spatial frequency domain with null points at $1/S$ ($S$ replaces $T$).

There are many reasons for converting from the time or space domain into the temporal or spatial frequency domain. Sometimes data are more meaningful in the frequency domain. For example, in the temporal frequency domain $\omega$, we can filter out specific frequencies that we consider noise, such as a single-frequency jammer. In the space domain, multiple plane waves from different directions (from distant sources) will strike a sensor array from different directions. The spatial Fourier transform will separate the plane waves into multiple discrete points in the spatial frequency domain (Section 14.1). A single-direction jammer can be removed in adaptive beamforming to prevent it from masking target information from a different direction. If a 2D transform is used with time $t$ on one axis and space $x$ (or angle) on the other, sources with different frequencies and directions will show up as different points on the 2D Fourier transform.

In image processing, we can remove spatially periodic background noise; for example, bars in front of a tiger in a cage can be filtered out in the spatial frequency domain. On taking the inverse transform, the tiger can be seen without the bars. Targets with spatial periodicity can be recognized, and early submarines could be identified by the periodic ribs in their structure.

Correlation and convolution are widely performed computations. The Fourier transform is normally used to perform convolution or correlation on a computer or in optics because convolution and correlation become multiplication in the frequency domain and the low cost of the fast Fourier transform in computers or the optical Fourier transform in optics makes the computation faster than a straight convolution or correlation computation [83]. Correlation is the basis for comparing images or searching for patterns or words in huge databases. Convolution implements linear systems such as smoothing, prediction, and many other signal processing functions. Chapter 14 of my previous book [83] describes an optical data flow computer that I designed as a PI to DARPA and that was later partially constructed for NSA. It performs ultrafast correlation.

## 3.2 UNCERTAINTY PRINCIPLE FOR FOURIER TRANSFORMS

Instead of computing the diffraction fields exactly, we can use the uncertainty principle for Fourier transforms to obtain a relation between the width of the field at the source and the smallest width of the information in the far field. The uncertainty principle for the Fourier transform is related to the Heisenberg uncertainty principle in quantum mechanics, $\Delta x \Delta p \geq \hbar/2$. This states that we cannot simultaneously determine the position of a particle ($\Delta x \to 0$) and its momentum ($\Delta p \to 0$) to arbitrarily high accuracy. We now show that a similar principle applies to Fourier transforms in time or space and how this may be used for a first approximation in lieu of performing a lengthy and detailed diffraction computation.

### 3.2.1 Uncertainty Principle for Fourier Transforms in Time

The uncertainty principle for Fourier transforms (or harmonics) in time specifies

$$\Delta t \Delta \omega \geq \frac{1}{2} \quad \text{or} \quad \Delta t \Delta v \geq \frac{1}{2} \tag{3.5}$$

where $\Delta t$ is the width in time and $\Delta v$ is the width in frequency, $\omega = 2\pi v$.

First, the reasonableness of the uncertainty principle, equation (3.5), can be seen from the Fourier transform scaling property [49, 133]. Consider the scaling property, if $\mathcal{F}\{g(t)\} = G(\omega)$, then

$$\mathcal{F}\{g(at)\} = \frac{1}{|a|} G\left(\frac{\omega}{a}\right) \tag{3.6}$$

Or if a pulse is stretched by $a$ in time to $at$, then the spectrum will become narrower by $1/a$ in frequency to $\omega/a$ and vice versa: the inverted scaling property of the Fourier transform. Consequently, we expect a limit to accurately predict both the location of the pulse and its frequency simultaneously. In the limit, a delta function in time, representing exact knowledge of time ($\Delta t \to 0$), has all frequencies equally, which prevents accurate determination of its frequency. Similarly, a single frequency, a sinusoid, represented by a delta function in frequency, giving exact knowledge of frequency ($\Delta \omega \to 0$), continues for all time, which prevents accurate measurement of its time of occurrence.

### 3.2.1.1  *Proof of Uncertainty Principle for Fourier Transforms in Time*

We note that the constant on the right-hand side of the equation (3.5) may vary depending on the definitions of width used for $\Delta t$ and $\Delta \omega$: possibilities are width at half maximum, root mean square, standard deviation, and Rayleigh width between nulls.

We follow the approach for complex functions $f$ used in Ref. [134]. A proof for real functions $f$ is given in Ref. [169].

$$E = \int_{-\infty}^{\infty} |f(t)|^2 \, dt, \qquad E = \frac{1}{2\pi} \int_{-\infty}^{\infty} |F(\omega)|^2 \, d\omega \tag{3.7}$$

Using Schwartz's inequality [134], $\left| \int p(t) q(t) dt \right|^2 \le \int |p(t)|^2 \, dt \int |q(t)|^2 dt$, with $p(t) = tf$ and $q(t) = df^*/dt$, we can write, reversing the direction of the equation,

$$\int_{-\infty}^{\infty} t^2 |f|^2 \, dt \int_{-\infty}^{\infty} \left| \frac{df}{dt} \right|^2 \, dt \ge \left| \int_{-\infty}^{\infty} (tf) \left( \frac{df^*}{dt} \right) \, dt \right|^2 \tag{3.8}$$

We show that the square root of the right-hand side of equation (3.8) leads to $E/2$, which is $E$ times the right-hand side of the uncertainty principle, equation (3.5); that is $\int_{-\infty}^{\infty} (tf)(df^*/dt) \, dt = E/2$. First, the square root of the right-hand side of equation (3.8) may be written as

$$\left| \int_{-\infty}^{\infty} tf \frac{df^*}{dt} \, dt \right| = \frac{1}{2} \left| \int_{-\infty}^{\infty} t \left( f \frac{df^*}{dt} + f^* \frac{df}{dt} \right) \, dt \right| \tag{3.9}$$

$$= \frac{1}{2} \left| \int_{-\infty}^{\infty} t \frac{d(ff^*)}{dt} dt \right| = \frac{1}{2} \left| \int_{-\infty}^{\infty} t \left( \frac{d|f|^2}{dt} \right) \, dt \right| = \frac{1}{2} \left| \int_{-\infty}^{\infty} t d |f|^2 \right|$$

$$= \frac{1}{2} \left( t|f|^2 \Big|_{-\infty}^{\infty} - \int_{-\infty}^{\infty} |f|^2 \, dt \right) = -\frac{E}{2} \tag{3.10}$$

where we used integration by parts and equation (3.7) for obtaining the last line. Also, as $t \to \infty$, $\sqrt{|t|} |f| \to 0 \Rightarrow t|f|^2 = 0$, so the first term of the last line in equation (3.10) is zero.

Now we show that the left-hand side of equation (3.8) leads to the product of width squared in time and width squared in frequency for the square of the uncertainty principle on the left-hand side of equation (3.5). Taking a derivative of a time function is equivalent to multiplying by $j\omega$ in the frequency domain and taking a second derivative is equivalent to multiplying by $\omega^2$ in the frequency domain. Therefore, if $F(\omega)$ is the Fourier transform of $f$, then

$$\int_{\infty}^{\infty} \left| \frac{\mathrm{d}f}{\mathrm{d}t} \right|^2 \mathrm{d}t = \frac{1}{2\pi} \int_{\infty}^{\infty} \omega^2 |F(\omega)|^2 \, \mathrm{d}\omega \tag{3.11}$$

Substituting equation (3.11) into the left-hand side of equation (3.8) and equation (3.10) into the right-hand side of equation (3.8) gives

$$\int_{-\infty}^{\infty} t^2 |f|^2 \, \mathrm{d}t \frac{1}{2\pi} \int_{-\infty}^{\infty} \omega^2 |F(\omega)|^2 \, \mathrm{d}\omega \geq \frac{E^2}{4} \tag{3.12}$$

If we define

$$(\Delta t)^2 = \frac{1}{E} \int_{-\infty}^{\infty} t^2 |f(t)|^2 \, \mathrm{d}t \quad \text{and} \quad (\Delta\omega)^2 = \frac{1}{E} \frac{1}{2\pi} \int_{-\infty}^{\infty} \omega^2 |F(\omega)|^2 \, \mathrm{d}\omega \tag{3.13}$$

then substituting equation (3.13) into equation (3.12) gives

$$(\Delta t)^2 (\Delta\omega)^2 \geq \frac{1}{4} \tag{3.14}$$

The inequality holds after taking square roots of both sides, so we get

$$\Delta t \Delta\omega \geq \frac{1}{2} \tag{3.15}$$

which proves the uncertainty principle for Fourier transforms, equation (3.5).

For equality to hold in equation (3.15), equality must also hold in equation (3.8) or $|\mathrm{d}f/\mathrm{d}t| = |kft|$. This occurs only for functions $f$ with a Gaussian magnitude because the Fourier transform of a Gaussian function is also Gaussian, $\exp\{-\pi(bt)^2\} \longleftrightarrow \exp\{-\pi(f/b)^2\}$.

In an alternative form, we scale equation (3.12) by changing variables to frequency $v$ in place of angular frequency $\omega = 2\pi v$ (and $\mathrm{d}\omega = 2\pi \, \mathrm{d}v$).

$$\int_{-\infty}^{\infty} t^2 |f|^2 \, \mathrm{d}t \int_{-\infty}^{\infty} v^2 |F(v)|^2 \, \mathrm{d}v \geq \frac{E^2}{4} \tag{3.16}$$

The uncertainty principle for Fourier transforms in time can then be written in terms of time and frequency widths as

$$\Delta t \Delta v \geq \frac{1}{2} \tag{3.17}$$

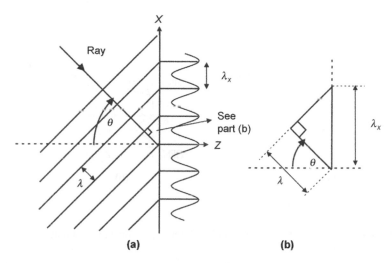

**FIGURE 3.3** Relation between beam angle $\theta$ and spatial frequency $\lambda_s$: (a) wave striking vertical screen and (b) triangle in (a).

### 3.2.2 Uncertainty Principle for Fourier Transforms in Space

Because of the interchangeability of space and time, we can derive an uncertainty principle for Fourier transforms in space in a similar manner. More simply, we replace time $t$ with space $s$ and temporal angular frequency $\omega = 2\pi\nu$ with spatial angular frequency $k_x = 2\pi f_x$ to obtain the equivalent uncertainty principle for Fourier transforms in space, the one that arises in diffraction. Note that space has up to three dimensions, unlike time. For simplicity, we consider only 1D.

$$\Delta s\Delta k_x \geq \frac{1}{2} \quad \text{or} \quad \Delta s\Delta f_x \geq \frac{1}{2} \tag{3.18}$$

The uncertainty principle may also be formulated in terms of emitted beam angle instead of spatial frequency. We consider a wave of wavelength $\lambda$ striking a plane at an angle $\theta$ with the normal (Figure 3.3a). The propagation vector $\mathbf{k}$ can be decomposed into $x$ and $z$ components according to $\mathbf{k}^2 = k_x^2 + k_z^2$. A wave component propagates down the screen in $x$ direction with propagation constant $k_x$ corresponding to a spatial frequency $f_x$ with wavelength $\lambda_x = 1/f_x$. Figure 3.3b shows a right triangle, extracted from Figure 3.3a, from which

$$\sin\theta = \frac{\lambda}{\lambda_x} = \lambda f_x \quad \text{for small angles} \quad \Delta f_x = \frac{\Delta\theta}{\lambda} \tag{3.19}$$

where for small angles, $\theta \to \Delta\theta$ and $\sin(\Delta\theta) \to \Delta\theta$. Substituting equation (3.19) into equation (3.18) gives an uncertainty principle for angle and distance:

$$\Delta s\Delta\theta \geq \frac{\lambda}{2} \tag{3.20}$$

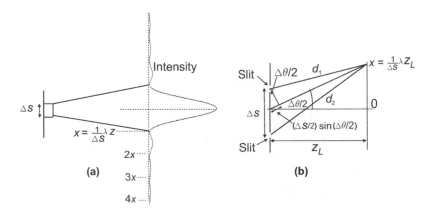

**FIGURE 3.4**  Fraunhofer diffraction as an interference phenomenon: (a) finite width source and (b) two slits.

In this case, it is not possible to make both the source size $\Delta s$ and the beam angle $\Delta\theta$ of light emission arbitrarily small. In fact, arrays in radar and sonar and beam expanders in optics increase $\Delta s$ to reduce beam angle in order to channel electromagnetic energy into a narrower beam for more accurate detection, imaging, or direction finding. A phased array radar or sonar scans this more piercing narrow beam electronically or it can be rotated. In the limit, if the source becomes infinitely small, it radiates in all directions and an angle cannot be distinguished. If, on the other hand, the beam angle is infinitely narrow, the source should be of infinite dimension.

### 3.2.2.1  *Relation Between Diffraction and Interference*  It is instructive to view diffraction from an interference perspective. We consider a plane wave front at a source of width $\Delta s$ (Figure 3.4a). A similar problem is presented by the two slits shown in Figure 3.4b. In either case, we consider two rays starting $\Delta s$ apart. At a screen at the Fraunhofer distance, the two beams will interfere to produce a spatial Fourier transform in $x$. A null occurs at point $P$ on the screen when the two beams arrive out of phase ($\pi$ phase difference) with each other for destructive interference. This corresponds to the difference in distance traveled by the beams, $d_2 - d_1 = \lambda/2$. For interference, we add the two beams, $E_1 + E_2$, and then compute their combined intensity, $\langle (E_1 + E_2) \rangle \langle (E_1 + E_2)^* \rangle$. As discussed in Chapter 6, if we assume that the two beams arrive at the target with equal power $A$, one of the form $\exp\{i(kd_1 - \omega t)\}$ and the other $\exp\{i(kd_2 - \omega t)\}$, the intensity of the field on the screen can be written as

$$I = 2A^2(1 + \cos(k_x \Delta d)) \qquad (3.21)$$

where $\Delta d = d_2 - d_1$. From Figure 3.4b, $\Delta d = \Delta s \sin(\Delta\theta/2) \approx \Delta sx/z_L$, so that equation (3.21) becomes, using $k_x = 2\pi/\lambda$,

$$I = 2A^2 \left[ (1 + \cos\left( \frac{2\pi \Delta sx}{\lambda z_L} \right) \right] \qquad (3.22)$$

where $\lambda z_L$ is characteristic of shrinkage when optical diffraction is used to compute a Fourier transform. For example, when $\lambda = 600\,\mathrm{nm}$ and distance to screen $z_L = 10\,\mathrm{cm}$, the spatial Fourier transform at the screen is smaller than would be calculated by multiplying by $\lambda z_L = 600 \times 10^{-9} \times 10^{-1} = 6 \times 10^{-8}$. The shrinkage is the basis of holographic optical memory [83] and espionage in which a page of data is shrunk to the size of a period in a letter for covert communication.

From equation (3.22), peaks or fringes occur on the $x$ axis at distances $x_f$ when $(2\pi \Delta s x_f)/(\lambda z_L) = 2\pi$ or $x_f = (\lambda z_L)/(\Delta s)$. As shown in time in Figure 3.4b, the distance to the first null from the axis is proportional to $1/\Delta s$. However, for space there is reduction by multiplication by $\lambda z_L$.

## 3.3 SCALAR DIFFRACTION

Most of us are familiar with reflection and refraction from mirrors and the apparent bend of a straw in water. The third interaction between coherent light and boundaries is diffraction. We are concerned with a monochromatic wave passing through a 2D mask and then propagating over a distance $z$ to a target. Given the electromagnetic wave field incident on the mask and the transmissive function $U(x, y)$ of the mask, we desire to know the electromagnetic field at the target. This will allow us to predict the field after a laser beam has propagated through the atmosphere to project energy either for a laser weapon (Section 12.2) or for transmitting or receiving signals in optical communications. It also allows us to modify a beam's wave front for adaptive optics (Sections 5.3 and 12.2) or determine the spot size for focused beams or Gaussian beams (Section 2.1) in order to concentrate power to high intensities (Section 14.2.2).

We restrict our mathematical development to scalar diffraction for simplicity; that is, we consider only one component of the electric field $E_z$ [49]. For the vector derivation, see Ref. [16]. Apparently, the scalar result provides satisfactory answers most of the time. Following Fresnel's award winning procedure, we break the derivation into four stages:

1. Find an expression for the field at a point due to the field on a boundary surrounding the point (Section 3.3.2).
2. Find the field behind an aperture (Section 3.3.3).
3. Use the Fresnel approximation for a linear system relating the field at point $P_0$ in the output plane to the field at point $P_1$ in the plane of the input mask for a plane wave incident on the mask and a distance limited to that involving space invariant transformation allowing convolution (Section 3.3.4).
4. Use the Fraunhofer (far-field) approximation to show the 2D Fourier transform relation between the field at $P_0$ and the field at $P_1$ (Section 3.3.5). A lens can bring the transform much closer [49, 83].

We note that the results may be useful after any step. We have used the result of item 1 to improve the magnetic detection of nuclear missile carrying submarines hiding their magnetic signature in shallow water for which the spatial variations of the earth's magnetic field are very large.

### 3.3.1    Preliminaries: Green's Function and Theorem

We select a spherically radiating field from a point source as Green's function,

$$G(r) = \frac{\exp\{jkr\}}{r} \tag{3.23}$$

where $r$ is the distance from the point source and $k = 2\pi/\lambda$ is the wave propagation constant in air. Here, we assume time harmonic (single-frequency) fields $\exp\{-j\omega t\}$. Note that some other texts assume $\exp\{j\omega t\}$, depending sometimes on whether the writer had a physics or electrical engineering background. $j$ and $i$ are similarly interchanged. In either case, the instantaneous field is obtained by multiplying the phasor used by this term and then taking the real part. Or an alternative that can be used for multiplied terms is to add the conjugate of the phasor to remove the imaginary part. A complicated source and its associated field can be constructed by summing many point sources.

We consider a field $U$, write the divergences of $G\nabla U$ and $U\nabla G$, apply the divergence theorem to both, and subtract one from the other to obtain Green's theorem [49]:

$$\int\!\!\int\!\!\int_v (G\nabla^2 U - U\nabla^2 G)\mathrm{d}v = \int\!\!\int_S \left(G\frac{\partial U}{\partial n} - U\frac{\partial G}{\partial n}\right)\mathrm{d}S \tag{3.24}$$

### 3.3.2    Field at a Point due to Field on a Boundary

To apply Green's theorem, equation (3.24), we require, at the boundary, the field $U$ and the normal field $\partial U/\partial n$ (Figure 3.5). As $U$ and $G$ satisfy the Helmholtz equation (the time harmonic wave equation) $(\nabla^2 + k^2)U = 0$,

$$\nabla^2 G = -k^2 G \quad \text{and} \quad \nabla^2 U = -k^2 U \tag{3.25}$$

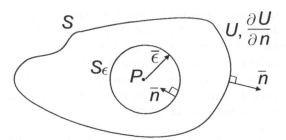

**FIGURE 3.5**    Finding field at point $P$ from that on a surrounding boundary.

By substituting equation (3.25), the left-hand side of equation (3.24) tends to zero. Furthermore, to avoid the difficulty of dealing with the field at a point (singularity), we draw a small sphere around the point $P$ and consider the surface of this sphere to be a second boundary $S_\epsilon$. $\epsilon$ will be allowed to shrink to zero later. Then,

$$0 = \int\int_{S_\epsilon} \left( G\frac{\partial U}{\partial n} - U\frac{\partial G}{\partial n} \right) dS_\epsilon + \int\int_S \left( G\frac{\partial U}{\partial n} - U\frac{\partial G}{\partial n} \right) dS \qquad (3.26)$$

Now consider Green's function $G = \exp\{jkr\}/r$. For the small sphere around $P$, $r = \epsilon$, so $G = \exp\{jk\epsilon\}/\epsilon$ and

$$\frac{\partial G}{\partial \epsilon} = \frac{1}{\epsilon} jk \exp\{jk\epsilon\} + \exp\{jk\epsilon\}\left(-\frac{1}{\epsilon^2}\right) = \frac{\exp\{jk\epsilon\}}{\epsilon}\left(jk - \frac{1}{\epsilon}\right) \qquad (3.27)$$

This should be modified as $\partial G/\partial n$ for equation (3.24):

$$\frac{\partial G}{\partial n} = \frac{\partial \epsilon}{\partial n}\frac{\partial G}{\partial \epsilon} = \cos(\bar{n}, \bar{\epsilon})\frac{\exp\{jk\epsilon\}}{\epsilon}\left(jk - \frac{1}{\epsilon}\right) \qquad (3.28)$$

where $a$ refers to a unit direction vector of magnitude $a$. As $\bar{n}$ is toward the center of small sphere and $\bar{\epsilon}$ is outward, $\cos(\bar{n}, \bar{\epsilon}) = -1$. Now substituting equation (3.28) into equation (3.26) gives

$$0 = \int\int_{S_\epsilon} \left[ \frac{\exp\{jk\epsilon\}}{\epsilon}\frac{\partial U}{\partial n} - U\frac{\exp\{jk\epsilon\}}{\epsilon}\left(\frac{1}{\epsilon} - jk\right) \right] dS_\epsilon$$
$$+ \int\int_S \left( G\frac{\partial U}{\partial n} - U\frac{\partial G}{\partial n} \right) dS \qquad (3.29)$$

For a point $P$, $\epsilon \to 0$, $dS_\epsilon \to 4\pi\epsilon^2$, $U \to U(P_0)$,

$$0 = 4\pi\epsilon^2\left[ \frac{\exp\{jk\epsilon\}}{\epsilon}\frac{\partial U(P_0)}{\partial n} - U(P_0)\frac{\exp\{jk\epsilon\}}{\epsilon}\left(\frac{1}{\epsilon} - jk\right) \right]$$
$$+ \int\int_S \left( G\frac{\partial U}{\partial n} - U\frac{\partial G}{\partial n} \right) dS \qquad (3.30)$$

Consider the first term of equation (3.30), $4\pi\epsilon^2 \times$ terms in the square bracket. When $\epsilon \to 0$, the only term remaining in the first term of equation (3.30) is $4\pi\epsilon^2 U(P_0)\exp\{jk\epsilon\}/\epsilon^2$. So

$$0 = -4\pi U(P_0) + \int\int_S \left( G\frac{\partial U}{\partial n} - U\frac{\partial G}{\partial n} \right) dS \qquad (3.31)$$

From this equation, the field $U(P_0)$ at a point $(P_0)$ inside a closed region for which the field $U$ and normal field $\partial U/\partial n$ are known along the boundary is

$$U(P_0) = \frac{1}{4\pi} \int \int_S \left( G \frac{\partial U}{\partial n} - U \frac{\partial G}{\partial n} \right) dS \tag{3.32}$$

### 3.3.3  Diffraction from an Aperture

We assume that the aperture through which light passes is placed into the boundary around the point at which we seek the field (Figure 3.6). First, we form a more convenient Green's function $G_2(P_1)$ that has two point sources of opposite phase so that for $G_2(P_1)$, the field at point $P_1$ cancels in the aperture:

$$G_2(P_1) = \frac{\exp\{jkr_{01}\}}{r_{01}} - \frac{\exp\{jkr_{02}\}}{r_{02}} \tag{3.33}$$

We assume that the second point source is located at a mirror image of the first around the plane of the aperture. Then, $r_{01} = r_{02}$ and from equation (3.33),

$$G_2(P_1) = 0 \tag{3.34}$$

Now using a similar process (using a bar over a variable to represent a unit vector) as for equation (3.28), but with $r_{01}$ replacing $\epsilon$,

$$\frac{\partial G_2}{\partial n} = \cos(\bar{n}, \bar{r}_{01}) \left( jk - \frac{1}{r_{01}} \right) \frac{\exp\{jkr_{01}\}}{r_{01}} - \cos(\bar{n}, \bar{r}_{02}) \left( jk - \frac{1}{r_{02}} \right) \frac{\exp\{jkr_{02}\}}{r_{02}} \tag{3.35}$$

As $\cos(\bar{n}, \bar{r}_{02}) = -\cos(\bar{n}, \bar{r}_{01})$, $k = 2\pi/\lambda$, and $r_{01} = r_{02}$

$$\frac{\partial G_2}{\partial n} = 2 \cos(\bar{n}, \bar{r}_{01}) \left( jk - \frac{1}{r_{01}} \right) \frac{\exp\{jkr_{01}\}}{r_{01}} \tag{3.36}$$

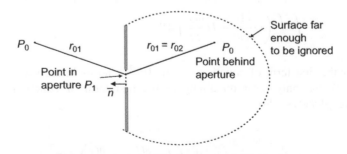

**FIGURE 3.6**  Diagram for deriving light behind an aperture.

If we assume that $r_{01} \gg \lambda$, $1/r_{01}$ may be neglected in $(jk - 1/r_{01})$, using $k = 2\pi/\lambda$,

$$\frac{\partial G_2}{\partial n} = -\frac{4\pi}{j\lambda} \frac{\exp\{jkr_{01}\}}{r_{01}} \cos(\bar{n}, \bar{r}_{01}) \tag{3.37}$$

For the field at point $P_0$, substitute equations (3.37) and (3.34) into equation (3.32). Label the surrounding surface $S_1$ and an arbitrary point in the aperture $P_1$:

$$U(P_0) = \frac{1}{4\pi} \int\int_{S_1} -U(P_1)\frac{\partial G_2}{\partial n} dS$$

$$= \frac{1}{j\lambda} \int\int_{S_1} U(P_1)\frac{\exp\{jkr_{01}\}}{r_{01}} \cos(\bar{n}, \bar{r}_{01}) dS \tag{3.38}$$

This gives the Rayleigh–Sommerfeld equation for an aperture, relabeled $\Sigma$,

$$U(P_0) = \int\int_{\Sigma} h(P_0, P_1)U(P_1)dS$$

$$\text{with} \quad h(P_0, P_1) = \frac{1}{j\lambda}\frac{\exp\{jkr_{01}\}}{r_{01}} \cos(\bar{n}, \bar{r}_{01}) \tag{3.39}$$

where $h(P_0, P_1)$ is the impulse response that describes a linear system representing diffraction during propagation.

We see how close Huygens came to the general structure in 1687 by rewriting equation (3.39):

$$U(P_0) = \int\int_{\Sigma} \overbrace{\frac{\exp\{jkr_{01}\}}{r_{01}}}^{\text{spherical wave}} \left[\overbrace{\frac{\cos(\bar{n}, \bar{r}_{01})}{\underbrace{j\lambda}_{\lambda \text{ dependence}}}}^{\text{angle loss}}\right] U(P_1)dS \tag{3.40}$$

Huygens, lacking different single frequency sources, ignored the square bracket in equation (3.40) in explaining a diffraction of a plane wave striking an aperture as regenerating a set of point sources for which each of the spherical waves emanated adds at a screen to produce an interference pattern (Figure 3.7).

### 3.3.4  Fresnel Approximation

We first assume that the distance $z$ between input mask and screen is much greater than the transverse dimension at the diffraction plane so that the obliquity factor $\cos(\bar{n}, \bar{r}_{01}) = 1$. In this case, $r_{01} \approx z$ in the denominator of equation (3.40). This cannot be used in the exponential phase term in the numerator because of the small size of wavelength $\lambda$. Now equation (3.39) can be written for propagation from

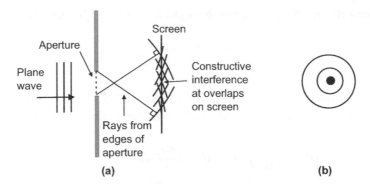

**FIGURE 3.7** Interpretation of diffraction as sum of fields from new point sources at an aperture: (a) spherical waves and (b) view of intensity on screen.

input plane $(x_1, y_1)$ to output plane $(x_0, y_0)$ (Figure 3.8) as

$$U(x_0, y_0) = \int\!\!\int_\Sigma h(x_0, y_0 : x_1, y_1)U(x_1, y_1)\mathrm{d}x_1\,\mathrm{d}y_1$$

$$\text{with} \quad h(x_0, y_0 : x_1, y_1) = \frac{1}{j\lambda}\frac{\exp\{jkr_{01}\}}{z} \qquad\qquad (3.41)$$

We now apply an approximation that will convert the linear system representing propagation with diffraction into a shift invariant system that is described by convolution. Convolution can be performed rapidly using the frequency domain and the convolution theorem of Fourier transform theory [49]. The spherical wave fronts generated by diffraction at the input mask, plane 1, are approximated by parabolas close to the axis because a parabola approximates a sphere over a small region; this is called the

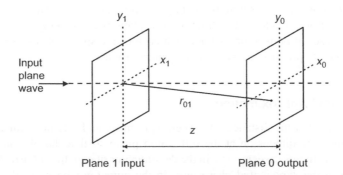

**FIGURE 3.8**   Diffraction from plane 1 to plane 0.

paraxial approximation (Section 1.1). Mathematically, we write

$$r_{01} = \sqrt{z^2 + (x_0 - x_1)^2 + (y_0 - y_1)^2}$$

$$= z\sqrt{1 + \left(\frac{x_0 - x_1}{z}\right)^2 + \left(\frac{y_0 - y_1}{z}\right)^2} \tag{3.42}$$

and apply the binomial theorem to remove the square root:

$$r_{01} = z + \frac{z}{2}\left(\frac{x_0 - x_1}{z}\right)^2 + \frac{z}{2}\left(\frac{y_0 - y_1}{z}\right)^2 \tag{3.43}$$

The $r_{01}$ in the exponent in equation (3.41) is replaced with equation (3.43) to give the impulse response for the diffraction to the Fresnel approximation distance:

$$h(x_0 - x_1, y_0 - y_1) \approx \frac{\exp\{jkz\}}{j\lambda z} \exp\left\{\frac{jk}{2z}\left[(x_0 - x_1)^2 + (y_0 - y_1)^2\right]\right\} \tag{3.44}$$

This result is shift invariant because output values $x_0$ and $y_0$ appear relative to only input values $x_1$ and $y_1$ as $x_0 - x_1$ and $y_0 - y_1$. Substituting equation (3.44) into equation (3.41) produces the output field $U(x_0, y_0)$ in terms of the input field $U(x_1, y_1)$:

$$U(x_0, y_0) = \frac{\exp\{jkz\}}{j\lambda z} \int\int_{-\infty}^{\infty} U(x_1, y_1)$$

$$\exp\left\{\frac{jk}{2z}\left[(x_0 - x_1)^2 + (y_0 - y_1)^2\right]\right\} dx_1\, dy_1 \tag{3.45}$$

Using the convolution symbol $\otimes$,

$$U(x_0, y_0) = U(x, y) \otimes h(x, y)$$

$$\text{with} \quad h(x, y) = \frac{\exp\{jkz\}}{j\lambda z} \exp\left\{\frac{jk}{2z}\left[(x_0 - x_1)^2 + (y_0 - y_1)^2\right]\right\} \tag{3.46}$$

For computation involving diffraction propagation, we frequently operate in the frequency domain. A Fourier transform pair [49] in $x$ and $y$

$$\mathcal{F}\left(\exp\{-\pi(a^2x^2 + b^2y^2)\}\right) = \frac{1}{|ab|} \exp\left[-\pi\left(\frac{f_x^2}{a^2} + \frac{f_y^2}{b^2}\right)\right] \tag{3.47}$$

is also Gaussian and can be used to compute the transfer function of the Gaussian impulse response, equation (3.46). In equation (3.46), we replace $jk/(2z)$ by $j\pi/(\lambda z)$ and by comparison with equation (3.47), we assign $-a^2 = -b^2 = j/(\lambda z)$. Then, as $1/(ab) = j\lambda z$, then [49].

$$H(f_x, f_y) = \mathcal{F}\{h(x, y)\} = \exp\{jkz\} \exp[-j\pi\lambda z(f_x^2 + f_y^2)] \qquad (3.48)$$

which is also Gaussian. In the frequency domain, the output after diffraction is the Fourier transform of the product of transfer function $H(f_x, f_y)$ and Fourier transform of the input (from the Fourier convolution theorem) [49]. Hence, the output image $U(x_0, y_0)$ at distances greater than the Fresnel distance may be computed by taking the 2D Fourier transform of the input $U(f_x, f_y)$, multiplying by the Fresnel diffraction transfer function $H(f_x, f_y)$, and taking the inverse Fourier transform, equation (3.49),

$$U(x_0, y_0) = \mathcal{F}^{-1}\left[U(f_x, f_y)H(f_x, f_y)\right] \qquad (3.49)$$

### 3.3.5 Fraunhofer Approximation

Figure 3.9 shows how the transverse profile of the light intensity varies with distance past an aperture. Close to the aperture, we have a shadow of the aperture in the shadow-casting region. As distance $z$ increases, we pass into the Fresnel zone described by convolution equations (3.46), in space, or (3.49), in spatial frequency. At greater distances, far field, we enter the Fraunhofer zone.

Expanding the squares and moving terms out of the integral that are not dependent on the integration variables, we can write the Fresnel–Kirchhoff equation,

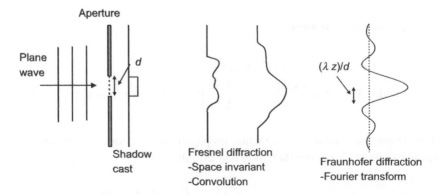

**FIGURE 3.9** Diffraction from plane 1 to plane of plane wave light through an aperture with distance.

equation (3.45), as

$$U(x_0, y_0) = \frac{\exp\{jkz\}}{j\lambda z} \exp\left[\frac{jk}{2z}(x_0^2 + y_0^2)\right] \int\int_{-\infty}^{\infty} U(x_1, y_1)$$

$$\overbrace{\exp\left[\frac{jk}{2z}(x_1^2 + y_1^2)\right]}^{q \text{ term}} \exp\left[-\frac{j\pi}{\lambda z}(x_0 x_1 + y_0 y_1)\right] dx_1\, dy_1 \quad (3.50)$$

where we used $k = 2\pi/\lambda$ or $k/(2z) = \pi/(\lambda z)$.

In the Fraunhofer zone, $z \gg k(x_1^2 + y_1^2)$ or $z \gg (2\pi/\lambda)(x_1^2 + y_1^2)$, or the distance $z$ of propagation is greater than the transverse distances relative to wavelength. In this case, the term labeled $q$ in equation (3.50) tends to one. Substituting this into equation (3.50) gives

$$U(x_0, y_0) = \frac{\exp\{jkz\}}{j\lambda z} \exp\left[\frac{jk}{2z}(x_0^2 + y_0^2)\right] \int\int_{-\infty}^{\infty} U(x_1, y_1)$$

$$\exp\left[-\frac{j2\pi}{\lambda z}(x_0 x_1 + y_0 y_1)\right] dx_1\, dy_1 \quad (3.51)$$

To show that the Fraunhofer output $U(x_0, y_0)$ is related to the scaled Fourier transform of the input, we scale the output by $\lambda z$ by defining a spatial frequency at the output of $f_x = x_0/(\lambda z)$ and $f_y = y_0/(\lambda z)$. Then we write equation (3.51) as

$$U(x_0, y_0) = \frac{\exp\{jkz\}}{j\lambda z} \exp\left[\frac{jk}{2z}(x_0^2 + y_0^2)\right] \int\int_{-\infty}^{\infty} U(x_1, y_1)$$

$$\exp\left[-j2\pi(f_x x_1 + f_y y_1)\right] dx_1\, dy_1 \quad (3.52)$$

Recognizing the integral as the spatial 2D Fourier transform of the input $U(x_1, y_1)$, we rewrite, including scaling, as

$$U(x_0, y_0) = \frac{\exp\{jkz\}}{j\lambda z} \exp\left[\frac{jk}{2z}(x_0^2 + y_0^2)\right] \mathcal{F}\{U(x_1, y_1)\}|_{f_x = x_0/(\lambda z), f_y = y_0/(\lambda z)}$$

$$(3.53)$$

The intensity of the output is

$$I(x_0, y_0) = |U(x_0, y_0)|^2 = \frac{1}{(\lambda z)^2} |\mathcal{F}\{U(x_1, y_1)\}|^2_{f_x = x_0/(\lambda z), f_y = y_0/(\lambda z)} \quad (3.54)$$

In summary, diffraction for distances greater than or equal to the Fraunhofer distance can be computed by taking the 2D Fourier transform of the input, scaling by replacing the spatial frequency with $f_x = x_0/(\lambda z)$ and $f_y = y_0/(\lambda z)$, and multiplying by prephase term $\exp\{jkz\}/(j\lambda z)\exp\{jk/(2z)(x_0^2 + y_0^2)\}$. The phase terms vanish for intensity.

### 3.3.6   Role of Numerical Computation

Propagation of physical waves, including optic waves, may be described mathematically and numerically by solutions to the wave equations that satisfy the boundary conditions and material properties. In normal mode theory, propagation is described by the sum of resonant modes, each of which is a solution to the wave equation. The combination of modes satisfies the boundary and source conditions. Mode theory is invaluable for analysis and when used in combination with algebraic computer programs (work with algebraic symbols in place of numbers) such as Maple and Mathematica. Mode theory can provide valuable information on relationships among parameters.

However, because the real world often has varying refractive index in space and boundaries are often too complicated for mode theory, we frequently solve the discretized wave equations directly with digital computers using finite approximation techniques. Discretizing the wave equation in both space and time is accomplished with finite-difference time-domain (FTDT) techniques [154]. FTDT provides insight into the distribution of energy in space and time as it propagates through inhomogeneous media. Discretizing in space only is accomplished with finite difference or finite elements [78, 79].

## 3.4   DIFFRACTION-LIMITED IMAGING

A circular lens captures light falling on the lens, so it not only performs a lens function but also acts as a circular aperture of diameter equal to that of the lens. In an imaging system, there will be a limiting aperture that limits the ability of the system to focus an input point source to an output point source. If the output is distorted only by the diffraction of the limiting aperture, the system is called a diffraction-limited system. A diffraction-limited system has no other aberrations and is therefore the best possible imaging system [49].

### 3.4.1   Intuitive Effect of Aperture in Imaging System

Figure 3.10 shows that the aperture of a convex lens loses high spatial frequencies in the focused output image of an input point source. Light from the input point source is expanded as a set of plane waves at different angles, the angular spectrum, as shown on the left-hand side of Figure 3.10. On striking a vertical plane at the lens, the source plane wave components at the widest angle from the axis correspond to those with the highest frequency at the plane of the lens. These fall outside the lens aperture so that they are lost in forming the output image. The loss of high spatial frequencies in passing through the aperture results in rings around the point source image and smoothing of the image. Smoothing arises because the high spatial frequencies provide the fine detail in an image. A larger diameter lens will capture higher spatial frequencies and the output image will appear closer to a point source, making the point sharper. This is the reason that a photographic camera will provide

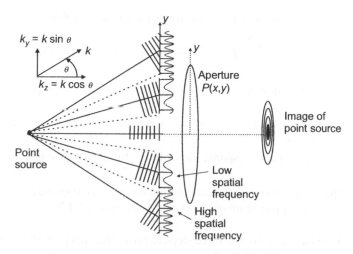

**FIGURE 3.10** Diffraction through aperture loses high spatial frequencies, causing smoothing of an image.

less sharp pictures with small apertures: high $f$-numbers. The large apertures used in dimmer light will also be less sharp but for a different reason: the edges of the lens are now used and these often have more aberrations than the center.

### 3.4.2 Computing the Diffraction Effect of a Lens Aperture on Imaging

We now show how to compute the effects of the aperture on an output image. The aperture is called a pupil function in the following. Figure 3.11 shows a simple imaging system with input object $U_0$ with coordinates $x_0$ and $y_0$ at distance $d_0$ from a lens. The lens with coordinates $x$ and $y$ has an optical field labeled $U_1$ in front of the lens and $U_1'$ after passing through the lens. An image $U_i$ with coordinates $x_i$ and $y_i$ is formed

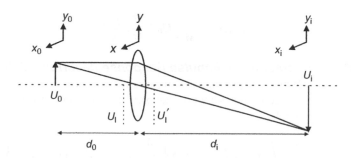

**FIGURE 3.11** Imaging a point source (at arrowhead) with a limited aperture lens.

at distance $d_i$ from the lens. Owing to linearity of wave propagation,

$$U_i(x_i, y_i) = \int_\infty^\infty \int_\infty^\infty h(x_i, y_i, x_o, y_o) U_o(x_o, y_o) dx_o \, dy_o \qquad (3.55)$$

The effects of diffraction on imaging can be computed in three steps:

1. Compute the geometric image by ignoring the effects of diffraction (Section 3.4.2.1).
2. Compute the impulse response describing diffraction through the system (Section 3.4.2.2).
3. For output image, compute the convolution between the impulse response from step 2 and the geometric image from step 1 (Section 3.4.2.3).

We proceed with equations for the three steps to prove that this procedure computes the correct image with diffraction effects.

### 3.4.2.1   Step 1: Compute the Geometric Image   We would like the image to be an identical copy of the input, except for scaling by $M$ to allow for magnification. This requires the impulse response from input to output to be a delta function of the form ($|M|^2$ is added to preserve total power)

$$h(x_i, y_i, x_o, y_o) = \frac{1}{|M|^2} \delta(x_i - Mx_o, y_i - My_o) \qquad (3.56)$$

The negatives account for image inversion through $M = -d_i/d_o = -x_i/x_o$. Substituting equation (3.56) into equation (3.55) shows that a scaled copy of the input is obtained:

$$U_i(x_i, y_i) = \int_\infty^\infty \int_\infty^\infty \frac{1}{|M|^2} \delta(x_i - Mx_o, y_i - My_o) U_o(x_o, y_o) dx_o \, dy_o \qquad (3.57)$$

From the sampling theorem for delta functions, the $\delta$ means sample $U_o$ at $x_i - Mx_o = 0$ or $x_o = x_i/M$ to give a geometric image that is a scaled version of the input:

$$U_g(x_i, y_i) = \frac{1}{|M|^2} U_o\left(\frac{x_i}{M}, \frac{y_i}{M}\right) \qquad (3.58)$$

### 3.4.2.2   Step 2: Compute the Impulse Response Describing Diffraction Through the System   The propagation impulse response from the output of the lens in Figure 3.11 to the output screen is adapted from Fresnel diffraction equation (3.45) (Section 3.3.4),

$$h(x_i, y_i, x, y) = \frac{1}{j\lambda d_i} \int \int_{-\infty}^\infty U_l'(x, y) \exp\left\{\frac{jk}{2d_i}\left[(x_i - x)^2 + (y_i - y)^2\right]\right\} dx \, dy \qquad (3.59)$$

where we neglected the leading phase term that is independent of $x$ and $y$.

Propagation through the lens with pupil $P(x, y)$ is represented by

$$U_1'(x, y) = U_1(x, y)P(x, y)\exp\left\{-\frac{jk}{2f}\left(x^2 + y^2\right)\right\} \tag{3.60}$$

Propagation from point source input object to the lens is represented by

$$U_1(x, y) = \frac{1}{j\lambda d_o}\exp\left\{\frac{jk}{2d_o}\left[(x - x_o)^2 + (y - y_o)^2\right]\right\} \tag{3.61}$$

Substituting equation (3.61) into equation (3.60) and the result into equation (3.59) gives the overall impulse response from the input point source to the output point source:

$$h(x_i, y_i, x_o, y_o)$$

$$= \frac{1}{\lambda^2 d_i d_o}\overbrace{\exp\left\{\frac{jk}{2d_o}\left(x_o^2 + y_o^2\right)\right\}}^{(a)}\overbrace{\exp\left\{\frac{jk}{2d_o}\left(x_i^2 + y_i^2\right)\right\}}^{(b)}$$

$$\int\int_{-\infty}^{\infty}P(x, y)\exp\left\{\frac{jk}{2}\overbrace{\left(\frac{1}{d_o} + \frac{1}{d_i} - \frac{1}{f}\right)}^{(c)}\left(x^2 + y^2\right)\right\}$$

$$\exp\left\{-jk\left[\left(\frac{x_o}{d_o} + \frac{x_i}{d_i}\right)x + \left(\frac{y_o}{d_o} + \frac{y_i}{d_i}\right)y\right]\right\}dx\,dy \tag{3.62}$$

For imaging, our interest is in magnitude; therefore, we ignore the phase terms (a) and (b). If we assume perfect imaging according to the lens law, equation (1.24), $1/d_o + 1/d_i - 1/f = 0$, term (c) is zero and as $\exp\{0\} = 1$, the exponential containing term (c) vanishes. We eliminate $d_o$ using equation (1.24) or $d_o = -d_i/M$. Then equation (3.62) becomes

$$h(x_i, y_i, x_o, y_o) = \frac{1}{\lambda^2 d_i d_o}$$

$$\int\int_{-\infty}^{\infty}P(x, y)\exp\left\{\frac{jk}{d_i}(x_i - Mx_o)x + (y_i - My_o)y\right\}dx\,dy \tag{3.63}$$

Equation (3.63) is not space invariant because of the magnification $M$. We force it to be space invariant, that is, in convolution form, by redefining variables $\hat{x}_o = |M|x_o$ and $\hat{y}_o = |M|y_o$. We also incorporate the scaling in new variables, $\hat{x} = x/(\lambda d_i)$ and $\hat{y} = y/(\lambda d_i)$ from which $dx = \lambda d_i\,d\hat{x}$ and $dy = \lambda d_i\,d\hat{y}$. In addition, using $k = 2\pi/\lambda$, equation (3.63) becomes (cancel $|M|$ from $1/(\lambda^2 d_1 d_o)$ and $P$)

$$h(x_i - \hat{x}_o, y_i - \hat{y}_o)$$

$$= \int\int_{-\infty}^{\infty}P(\lambda d_i\hat{x}, \lambda d_i\hat{y})\exp\left\{-j2\pi\left[(x_i - \hat{x}_o)\hat{x} + (y_i - \hat{y}_o)\hat{y}\right]\right\}d\hat{x}\,d\hat{y} \tag{3.64}$$

Equation (3.64) is the Fraunhofer diffraction of the aperture; note the scaling by $\lambda d_i$ in the pupil function for Fraunhofer.

### 3.4.2.3 Step 3. Show That Output Image is Convolution of Geometric Image and Fraunhofer Diffraction of Aperture Function: The output image may be written (convolution version of equation (3.55)) as

$$U_i(x_i, y_i) = \int_\infty^\infty \int_\infty^\infty h(x_i - \hat{x}_0, y_i - \hat{y}_0)U_0(x_0, y_0)dx_0\,dy_0 \qquad (3.65)$$

From text preceding equation (3.64), $\hat{x}_0 = |M|x_0$ and $\hat{y}_0 = |M|y_0$, so $dx_0 = d\hat{x}_0/|M|$ and $dy_0 = d\hat{y}_0/|M|$ in equation (3.65) give

$$U_i(x_i, y_i) = \int_\infty^\infty \int_\infty^\infty h(x_i - \hat{x}_0, y_i - \hat{y}_0)U_0\left(\frac{\hat{x}_0}{M}, \frac{\hat{y}_0}{M}\right)\frac{1}{|M|}\frac{1}{|M|}d\hat{x}_0\,d\hat{y}_0 \qquad (3.66)$$

From equation (3.58),

$$U_0\left(\frac{\hat{x}_0}{M}, \frac{\hat{y}_0}{M}\right) = |M|^2 U_g(\hat{x}_0, \hat{y}_0) \qquad (3.67)$$

Substituting $U_0(\hat{x}_0/M, \hat{y}_0/M)$ from equation (3.67) into equation (3.66) gives

$$U_i(x_i, y_i) = \int_\infty^\infty \int_\infty^\infty h(x_i - \hat{x}_0, y_i - \hat{y}_0)U_g(\hat{x}_0, \hat{y}_0)\,d\hat{x}_0\,d\hat{y}_0 \qquad (3.68)$$

The final result for imaging limited by diffraction of an aperture, equation (3.68), is the convolution of the geometric image, $U_g(x, y)$, with the Fraunhofer diffraction of the pupil (or aperture) function, $h(x, y)$ (scaled by $\lambda d_i$), equation (3.64),

$$U_i(x, y) = h(x, y) * U_g(x, y) \qquad (3.69)$$

For a square aperture, the pupil function has rectangular transmission function in $x$ or $y$. The aperture is scaled by $\lambda d_i$ and then Fourier transformed to find the Fraunhofer diffraction of the aperture. It will have the shape shown in Figure 3.2b. Convolving with the geometric image will smooth the image out, that is, remove high spatial frequencies. The Fraunhofer diffraction function is wider for smaller apertures, by the Fourier transform scaling property, so that smaller apertures will provide more smoothing.

# CHAPTER 4

# DIFFRACTIVE OPTICAL ELEMENTS

Diffractive optical elements (DOEs) manipulate light by diffraction (Chapter 3), rather than by refraction with lenses or reflection with mirrors. DOEs often consist of a thin piece of transparent material with an opaque or etched pattern that diffracts light into the desired form. The different behavior and manufacturing of DOEs relative to refractive elements has led to a significant diffractive optic element industry [69, 152] to complement the refractive lens industry (or mirror equivalents, Section 1.3).

The diffractive optical elements are based on the theory developed in Chapter 3, where we derived the equations for free-space Fresnel and Fraunhofer diffraction. For high-power optics (Section 12.2), light is manipulated with DOEs and dielectric mirrors (Section 6.3) to avoid conductors that overheat.

In Section 4.1, we briefly discuss some important applications of DOEs. In Section 4.2, we describe gratings for bending light and for separating colors or wavelengths for spectrometry in chemical analysis (Chapter 15) and wavelength division multiplexing (WDM). The grating equations are similar to those for an antenna array. Gratings can be created without lenses in the Talbot effect [85, 111]. In Section 4.3, we design a zone plate to replace a convex lens. In Section 4.4, we describe a widely used algorithm, Gerchberg–Saxton algorithm, for computing a phase-only diffractive element to perform 2D filtering without loss.

*Military Laser Technology for Defense: Technology for Revolutionizing 21st Century Warfare*, First Edition. By Alastair D. McAulay.
© 2011 John Wiley & Sons, Inc. Published 2011 by John Wiley & Sons, Inc.

## 4.1   APPLICATIONS OF DOEs

1. *Optical Elements for High Optical Power at Specified Wavelengths*: DOEs are often used with high-power laser beams in defense (Section 12.2) and manufacturing, welding, and cutting, where the optical element materials need to be able to withstand the power at the specified wavelength. More robust heat resisting and conducting materials, such as diamond and zinc selenide, are required for burning with infrared than are required at other wavelengths (Section 8.1.5). With these materials, it is often easier to make diffractive elements such as a focusing zone plate by machining or etching than a refractive lens by grinding and polishing.

2. *Light Weight*: DOEs may be used where light weight and low inertia are critical, such as in airborne platforms. In a compact disk player, replacing the laser diode lens with a lightweight focusing zone plate DOE can reduce switching time between channels. The collimating lens in an overhead transparency projector is designed by collapsing rings of the appropriate lens into a thin light molded plastic sheet.

3. *Complex Light Patterns*: Unlike refractive elements, DOEs can manipulate light into arbitrary complex beam patterns. For example, in supermarket scanner, a DOE produces multiple direction beams from one beam for reading bar codes at any angle.

4. *Spectroscopic Instruments*: DOEs, often gratings, can separate frequencies for spectroscopy or merge frequencies together (Chapter 15).

5. *Diffractive and Refractive Combinations*: Diffractive elements bend high frequencies more than low frequencies, which is opposite to the behavior of high-quality conventional spherical lenses that are easily made by grinding and polishing. Hence, putting a diffractive pattern such as a zone plate on a lens can cause it to focus all frequencies to the same spot. This is useful for surgery where the focused cutting beam is invisible IR and a focusing guide beam is visible. Molded lenses can be made with any shape; for example, aspheric lenses have elliptical or hyperbolic shape and will focus all colors to the same spot. But molded lenses, mostly plastic, are of lower quality and cannot be polished to high quality. An alternative to molding is to use the unusual process of sol-gels (melting glass powder at very high temperature).

## 4.2   DIFFRACTION GRATINGS

Gratings [49] are periodic structures used extensively in military laser systems: waveguide filters, optical fiber filters, communications, and spectrum analyzers (Chapter 15). Periodic structures [63, 147, 176] in more dimensions, photonic crystals and photonic crystal fibers [14], are important for future integrated optics. Figure 4.1 shows sinusoidal (a) and blazed (b) gratings. The grating is typically constructed by ruling or etching (by lithography) lines on a smooth glass substrate or by holographic

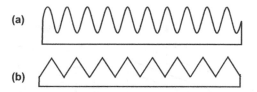

(a)

(b)

**FIGURE 4.1**   Gratings: (a) sinusoidal and (b) blazed.

techniques (for sinusoidal gratings) [83]. The grating can be transmissive or reflective for which it is coated with a reflecting surface. For a reflective grating, the blazed grating emphasizes certain reflection orders relative to the sinusoidal grating; therefore, these orders have a higher reflectance and are more efficient.

### 4.2.1   Bending Light with Diffraction Gratings and Grating Equation

A beam of light can be bent by both diffraction (Figure 4.2c) and refraction (Figure 4.2a and b). In Figure 4.2a, a beam from left strikes an interface with a material of lower refractive index so that the side of the beam that strikes the interface first is speeded up causing the beam to bend, similar to the bending of a squad of soldiers marching at an angle out of a muddy field into a parade ground. This explains the trapping of light in optical fiber by total internal reflection. The reverse direction for the beam also shows bending. The amount of bend is specified by Snell's law (Section 1.2.2). In Figure 4.2b, a prism has a longer path in glass for one side of the beam than the other side, which slows down that side causing the beam to bend.

In Figure 4.2c, bending by diffraction, one side of the beam, labeled $C$, travels farther than the other side $A$ if a bent plane wave front is desired at the output. Hence, the slowing, of the light of one side of the beam causes the beam to bend and for a grating period of $d$, the extra distance traveled is $d \sin \theta_1 + d \sin \theta_2$. For the output to have a plane wave front as shown, the extra distance traveled must be an integer multiple $m$ of a wavelength $\lambda$. This gives a grating equation

$$d \sin \theta_1 + d \sin \theta_2 = m\lambda \qquad (4.1)$$

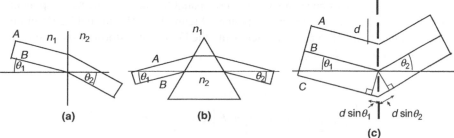

(a)                          (b)                          (c)

**FIGURE 4.2**   Bending a light beam: (a) by an interface with different refractive indices by refraction, (b) by a prism by refraction, and (c) by a grating by diffraction.

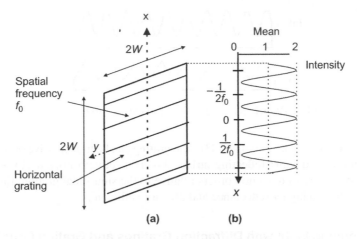

**FIGURE 4.3**   Cosinusoidal grating shape.

A major difference between diffractive and refractive bending (Figure 4.2c and a or b, is that in diffractive elements long wavelengths such as infrared make the grating look finer and are bent more than shorter wavelengths, while in refractive elements the reverse occurs; that is, the shorter wavelengths are bent more.

### 4.2.2   Cosinusoidal Grating

Figure 4.3a shows a cosinusoidal transmission grating of size $2W \times 2W$ and grating frequency $f_0$ for which the cross-sectional transmittance profile in $x$ (Figure 4.3b) may be written as

$$U_{\text{in}}(x, y) = \left[ \frac{1}{2} + \frac{1}{2} \cos(2\pi f_0 x_1) \right] \text{rect}\left( \frac{x_0}{2W} \right) \text{rect}\left( \frac{y_0}{2W} \right) \qquad (4.2)$$

A blazed grating is analyzed in Section 15.3.3. The first $\frac{1}{2}$ in equation (4.2) arises because intensity cannot be negative, so a zero spatial frequency or a DC component exists to produce the zero[th]-order component (Figure 4.4). The far-field diffraction (called Fraunhofer field) may be written using the Fourier transform of the intensity immediately after the grating:

$$\begin{aligned}
U_{\text{out}}(x_0, y_0) &= \mathcal{F}\{U_{\text{in}}(x_1, y_1)\} \\
&= \mathcal{F}\left[ \frac{1}{2} + \frac{1}{2}\cos(2\pi f_0 x_1) \right] * \mathcal{F}\left[ \text{rect}\left( \frac{x_0}{2W} \right) \text{rect}\left( \frac{x_0}{2W} \right) \right] \\
&= \left[ \frac{1}{2}\delta(f_x f_y) + \frac{1}{4}\delta(f_x + f_0, f_y) + \frac{1}{4}\delta(f_x - f_0, f_y) \right] \\
&\quad * (2W)^2 \text{sinc}(2W f_x)\text{sinc}(2W f_y)
\end{aligned}$$

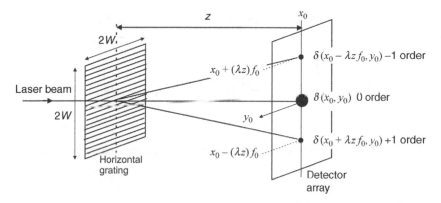

**FIGURE 4.4**  Fraunhofer field from a cosinusoidal grating.

$$= \frac{1}{2}(2W)^2\text{sinc}(2Wf_y)\left[\text{sinc}(2Wf_x) + \frac{1}{2}\text{sinc}\{2W(f_x + f_0)\}\right.$$
$$\left. + \frac{1}{2}\text{sinc}\{2W(f_x - f_0)\}\right] \quad (4.3)$$

where we used the Fourier transform property that the convolution in one domain becomes multiplication in the other and $\cos\theta = (\exp\{j\theta\} + \exp\{-j\theta\})/2$. The Fraunhofer diffraction field may be written from equation (3.53) (Section 3.3.5), using the Fourier transform from equation (4.3) and scaling with $f_x = x_0/(\lambda z)$ and $f_y = y_0/(\lambda z)$:

$$U(x_0, y_0) = \frac{\exp(jkz)}{j\lambda z}\exp\left[\frac{jk}{2z}\{x_0^2 + y_0^2\}\right]\frac{1}{2}(2W^2)\text{sinc}\left(\frac{2Wy_0}{\lambda z}\right)$$
$$\left[\text{sinc}\left(\frac{2Wx_0}{\lambda z}\right) + \frac{1}{2}\text{sinc}\left\{\frac{2W}{\lambda z}(x_0 + \lambda z f_0)\right\}\right.$$
$$\left. + \frac{1}{2}\text{sinc}\left\{\frac{2W}{\lambda z}(x_0 - \lambda z f_0)\right\}\right] \quad (4.4)$$

The three orders in the square brackets correspond with those shown in Figure 4.4. The far-field shape in the $x$ direction from a cosinusoidal grating, equation (4.4), at a far-field plane is plotted in Figure 4.5. For sufficient spacing to avoid overlap, as shown in Figure 4.4, of diffraction orders, we can write far-field intensity at the output by multiplying by its conjugate:

$$I_{\text{out}}(x_0 y_0) = \frac{1}{(\lambda z)^2}\left\{\frac{1}{2}(2W)^2\right\}^2\text{sinc}^2\left(\frac{2Wy_0}{\lambda z}\right)$$
$$\left[\text{sinc}^2\left(\frac{2Wx_0}{\lambda z}\right) + \frac{1}{4}\text{sinc}^2\left\{\frac{2W}{\lambda z}(x_0 + \lambda z f_0)\right\}\right.$$
$$\left. + \frac{1}{4}\text{sinc}^2\left\{\frac{2W}{\lambda z}(x_0 - \lambda z f_0)\right\}\right] \quad (4.5)$$

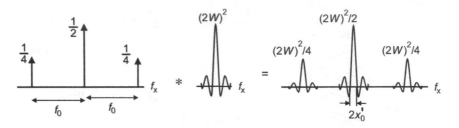

**FIGURE 4.5**   Far field from a cosinusoidal grating shows the three orders and their shape.

### 4.2.3   Performance of Grating

The separation of the $-1$ and $+1$ orders from the zeroth order, the direction the light would travel in the absence of the grating, is proportional to the spatial frequency $f_0$ of the grating. For diffraction in optics, the output is scaled by $\lambda z$, the wavelength $\lambda$ and the distance to the output plane $z$ (Figure 4.4), equation (4.5). Note that the more dense the grating, that is, lines close to each other, the larger the $f_0$, and the diffraction orders spread more. Also, the shorter the wavelength of the light, (toward UV), the smaller the $\lambda$, and the diffraction orders are bent farther away from the zeroth order. This is used for spectral analysis in Chapter 15 to identify chemical and biological weapons. As mentioned earlier, light is bent in the direction opposite to that in a prism, making diffractive–refractive spherical elements possible that, for example, will focus different colors to the same position, useful for eye surgery or weapon systems where burning is conducted with infrared while a tracking red light is observed. An aspheric lens to perform this function without a diffractive element cannot be made with commonly used spherical lens grinding systems.

The half-width of the diffraction order $x_0'$ seen in Figure 4.5 is obtained by setting the sinc term to zero in equation (4.4):

$$\text{sinc}\left(\frac{2W}{\lambda z}x_0'\right) = \frac{\sin\left(\pi 2Wx_0'/\lambda z\right)}{\pi 2Wx_0'/\lambda z} = 0 \qquad (4.6)$$

which occurs for

$$\pi\frac{2W}{\lambda z}x_0' = \pi \quad \text{or} \quad x_0' = \frac{\lambda z}{2W} \qquad (4.7)$$

Equation (4.7) shows that the width becomes narrower as the grating width $2W$ is increased, an anticipated outcome from Fourier transform inverse relations between output and input. The scaling $\lambda z$ indicates that width increases with distance and longer wavelengths.

## 4.3 ZONE PLATE DESIGN AND SIMULATION

A zone plate is a widely used diffractive optical element that performs a function similar to a refractive lens, except that it is manufactured differently and separates different wavelengths in opposite directions. It is used for light powers that would damage a normal refractive lens, or when a lens profile is required that is impossible to make by grinding with a spherical tool, or where an especially light lens is required. The equations are derived for a zone plate and for the field profile at the focal plane for the first-order diffraction. For a selected focal length, the radii and the etch depth, in the case of a surface relief zone plate, are computed for the purpose of fabrication. The transverse intensity profile and the peak field value on axis are computed and compared with a refractive lens.

### 4.3.1 Appearance and Focusing of Zone Plate

For a zone plate, we describe its appearance and how it focuses light.

***4.3.1.1 Appearance of Zone Plate*** A zone plate is defined by concentric rings of increasing radii with decreasing separation between rings with radius [61, 86, 134]. There are two types of zone plates: blocking and surface relief. In the blocking zone plate (Figure 4.6a), alternate rings are opaque to block light. This is also the appearance of the scaled mask for fabricating the surface relief type of zone plate. Note that most lithography machines reduce the mask size by 10 in printing the zone plate. In the surface relief zone plate, the opaque regions, instead of being opaque, are raised or lowered by an integer multiple of half-wavelengths in the material. A 3D view of the relief zone plate is shown in Figure 4.6b and a section through it is shown in Figure 4.7.

***4.3.1.2 Focusing of Zone Plate*** A zone plate focuses light by diffraction rather than by refraction as in a lens. Figure 4.8 shows the cross section of a blocking zone plate and how the zone plate focuses a parallel beam of light onto a spot at the

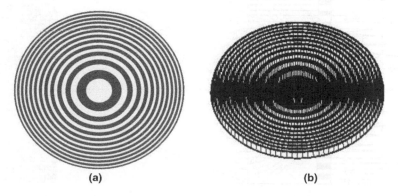

(a)          (b)

**FIGURE 4.6** View zone plates: (a) a blocking zone plate (and scaled mask for any zone plate) (b) a 3D view of a surface relief zone plate.

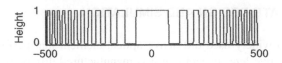

**FIGURE 4.7**   Cross section of 3D surface relief zone plate.

output plane. An equivalent focal length $f_p$ for the zone plate corresponds to the focal length $f$ of a lens. The zone plate may be viewed as a grating that becomes finer with radius out from the axis. A finer grating bends light more (Section 4.2.1), so light farther from the axis at the zone plate is bent more, causing an incoming parallel beam to be focused like a lens.

From another point of view, the distance from points on the zone plate to the focus point varies with radial distance from the axis in the zone plate; hence, the rays arrive with varying phase as a function of radius. The phase for each ray can be determined by the geometric length. For a blocking zone plate, those rays that are in phase at the focal point are passed through clear rings of the DOE. For a blocking zone plate, those rays that would be out of phase at the focal point are blocked by opaque rings of the DOE (Figure 4.8). For a surface relief plate, the intensity at the focal point is increased because phase interference replaces blocking of light [61]. Different colors and diffraction orders focus to different points along the axis. We consider a single wavelength and first-order diffraction. If the zone plate has high diffraction efficiency, we can arrange for negligible light to remain in the zero-order (or straight-through) diffraction.

### 4.3.2   Zone Plate Computation for Design and Simulation

The radii for the rings in the zone plate are computed to give a specified focal length.

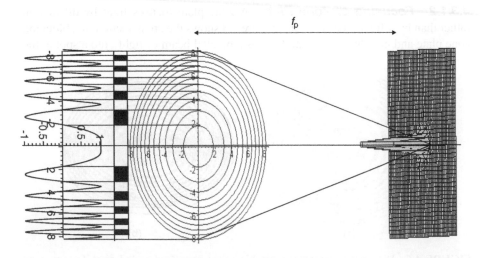

**FIGURE 4.8**   Focusing of zone plate.

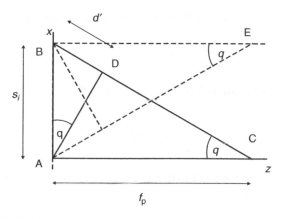

**FIGURE 4.9** Geometry for determining zone plate equations.

### 4.3.2.1 Derivation of Equations for Zone Plate Design

For the blocking zone plate (Figure 4.6a), in-phase rays at the focal point are passed through clear rings of the DOE. Those rays that would be out of phase at the focal point are blocked by opaque rings. The phase for each ray can be determined by its geometric length. The equations can be extended to surface relief zone plate (Figure 4.6b) by allowing a $\pi$ phase change in place of the blocked or opaque regions. This is equivalent to a change in sign, so that the previously out-of-phase parts are now in phase. This is normally used in diffractive elements to increase efficiency. The resulting intensities obtained here are doubled in this case. A $\pi$ phase change is achieved in a surface relief zone plate by exact control of the etch depth so that propagation through this depth of material will cause a $\pi$ phase change relative to propagation through the same depth in air.

We now compute the radii $s_i$ for the rings of a zone plate that focuses at a focal length of $f_p$ (Figure 4.8) [134]. Figure 4.9 shows at B the edge of the $i$th ring at distance $s_i$ from the axis for the zone plate focusing to a point C at the equivalent focal length on axis $f_p$. For the ray from the $i$th ring at radius $s_i$ to be at a transition between in phase and out of phase relative to that from the on-axis center of the zone plate, the additional distance traveled (XD) to the focal point is an odd number of half-wavelengths:

$$d' = \frac{\lambda}{2}(2i - 1) \tag{4.8}$$

From equation (4.8), similar triangles $ABC$ and $DBA$, $\lambda = (2\pi)/k$, and $AD \approx AB$,

$$\frac{s_i}{f_p} = \frac{(\pi/k)(2i - 1)}{s_i} = \frac{\pi(2i - 1)}{ks_i} \tag{4.9}$$

We select an equivalent focal length for the zone plate and define a parameter

$$\sigma^2 = f_p/k \tag{4.10}$$

Substituting $f_p$ from equation (4.10) into equation (4.9),

$$\frac{s_i^2}{\sigma^2} = \pi(2i - 1) \tag{4.11}$$

Therefore, opaque rings can be placed at intervals $s_{2i} \leq \rho \leq s_{2i+1}$, where the radius is $\rho$, $s_0 = 0$, and integers $i > 0$:

$$s_i = \sigma\sqrt{\pi(2i - 1)} \tag{4.12}$$

The opaque rings are marked black on the cross section at the left-hand side of Figure 4.8. Because the zeros of $U[\cos\{\rho^2/(2\sigma^2)\}]$ occur when $\rho = s_i$ in equation (4.12), the transmission function for the zone plate may be written as

$$T(\rho) = U\left[\cos\left(\frac{\rho^2}{2\sigma^2}\right)\right], \qquad U(x) = \begin{cases} 1 & \text{for } x \geq 0 \\ 0 & \text{for } x < 0 \end{cases} \tag{4.13}$$

Equation (4.13) is used to compute the cross section for the zone plate.

### 4.3.2.2 Derivation of Equations for Transverse Field at Focal Plane
To find the intensity of the field at the output plane, we use scalar diffraction theory (Section 3.3). Fresnel diffraction can be written from equation (3.45) as

$$g(x_0, y_0, z_0) = -\frac{j}{\lambda z_0}e^{jkz_0}\int_y\int_x f(x, y)\exp\left\{j\frac{k}{2z_0}\left[(x - x_0)^2 + (y - y_0)^2\right]\right\}dx\,dy \tag{4.14}$$

For circular symmetry $\rho = \sqrt{x^2 + y^2}$, equation (4.14) can be written using a Fourier–Bessel transform [49, 134] as

$$g(\rho_0, z_0) = -\frac{j2\pi}{\lambda z_0}e^{jkr_0}\int_0^b \rho f(\rho)\exp\left\{j\frac{k\rho^2}{2z_0}\right\}J_0\left(\frac{k\rho\rho_0}{z_0}\right)d\rho \tag{4.15}$$

where $r_0 = z_0 + (x_0^2 + y_0^2)/(2z_0)$ and $J_0$ is a Bessel function of first kind and zeroth order.

Inserting the zone plate design equations (4.13) and (4.10) into equation (4.15) in place of $f(\rho)$, $B$ is the amplitude of the incoming wave, and by changing variables using $w = k\rho^2/(2z_0)$ $(d\omega = (k\rho)/(z_0)d\rho)$ or $\rho^2 = (2z_0\omega)/k$, we obtain [134]

$$g(\rho_0, z_0) = B \int_0^{kb^2/(2z_0)} e^{jw} U\left[\cos\left(\frac{wz_0}{f}\right)\right] J_0\left[\rho_0\left(\frac{2kw}{z_0}\right)^{1/2}\right] dw \qquad (4.16)$$

where the phase term $-je^{jkr_0}$ is neglected. From equation (4.16), for the zone plate with $n$ rings and transparent ring locations (left-hand side of Figure 4.8),

$$g(\rho_0, z_0) \qquad\qquad\qquad\qquad\qquad\qquad\qquad (4.17)$$

$$= B \int_0^{\pi/2} e^{jw} J_0\left[\rho_0\left(\frac{2kw}{z_0}\right)^{1/2}\right] dw$$

$$+ B \int_{3\pi/2}^{5\pi/2} e^{jw} J_0\left[\rho_0\left(\frac{2kw}{z_0}\right)^{1/2}\right] dw + \cdots$$

$$+ B \int_{\pi(n-3/2)}^{\pi(n-1/2)} e^{jw} J_0\left[\rho_0\left(\frac{2kw}{z_0}\right)^{1/2}\right] dw \qquad (4.18)$$

### 4.3.2.3 *Performance of Zone Plate*   On the $z$ axis, $\rho_0 = 0$, and as $J_0 = 1$, the field at the focal point of the zone plate for $n$ rings is from equation (4.18)

$$g(\rho_0, z_0) = B\left[[1+j] + [1+1] + \cdots + [1+1]\right] = B[n+j]$$

Equation (4.18) shows that incoming light is focused to a higher intensity at the focal point.

The solid line in Figure 4.10 shows the normalized output intensity at the back focal plane for the zone plate illuminated with collimated coherent light. Note that

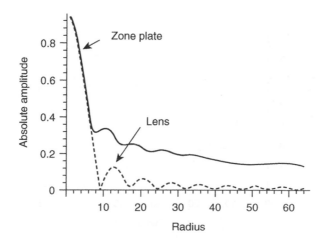

**FIGURE 4.10**   Comparison of profile at focal plane with a lens.

the zone plate will divert light into many higher orders because it is binary. Therefore, there are other higher order focus points at greater distances than the first order. Light in negative orders is spread around a circle and not focused at a point, so its peak intensity is significantly less than that at the focal point of a lens.

The field across the focal plane, $z_0 = f$, for a lens can be obtained from equation (4.15) by allowing the lens function to cancel the first exponent in the integral and setting $f(\rho) = 1$. By changing variables to $\rho' = k\rho\rho_0/z_0$ and using $\int_0^x \rho' J_0(\rho')d\rho' = xJ_1(x)$, this can be written as

$$g(\rho_0, z_0) = -\frac{j2\pi}{\lambda f}e^{jkr_0}b\frac{J_1(bk\rho_0/f)}{k\rho_0/f} \qquad (4.19)$$

As $J_1(bk\rho_0/f_p) = 0$, when $bk\rho_0/f_p = 1.22\pi$, the distance $2\rho_s$ across the main lobe, or the spot size, is

$$2\rho_s = \frac{2.44\lambda f}{2b} \qquad (4.20)$$

We consider the surface relief zone plate. If the opaque blocking regions are replaced by the material of the zone plate with additional height or depth such that $180°$ ($\pi$) phase shift occurs in these regions, then we add the same intensity again for these regions. The intensity at the peak for the surface relief zone plate is therefore twice that for the blocking zone plate. The intensity of the spot at the focal length will decrease if the etch depth deviates from that computed for $180°$ phase shift. Figure 4.11 shows the computed profile, rescaled, at the focal plane for a surface relief zone plate: This represents a significant improvement over the blocking zone plate shown previously because the side lobes fall off faster.

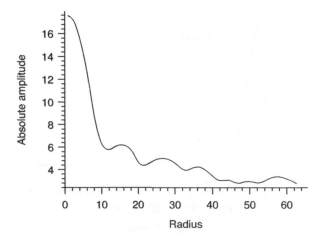

**FIGURE 4.11**   Computed profile at the focal plane for a surface relief zone plate.

## 4.4 GERCHBERG–SAXTON ALGORITHM FOR DESIGN OF DOEs

The Gerchberg–Saxton algorithm [43, 69, 152], is an error reduction algorithm to design diffractive optical elements that are phase only, so that the holographic-like single film DOE does not block any light [49, 83].

### 4.4.1 Goal of Gerchberg–Saxton Algorithm

Figure 4.12 shows how a (phase-only) diffractive element may be used at an input plane in a system to manipulate light from a plane-wave source, $W_0$ to a desired magnitude $|F_0(x, y)| = B_0(x, y)$ dependent of $x$ and $y$ at an observation plane that is distance $z$ from the filter plane. The goal of the adaptive iterative Gerchberg–Saxton algorithm is to compute the phase-only values as a function of $x$ and $y$ to diffract an input beam into a desired amplitude pattern at the output plane. This algorithm is used to design filters that can be placed after the laser diode in a laser pointer so that an image of an arrow or other object will appear on the screen in place of a spot. We have also used this algorithm for the design of a 2D phased array antenna, selecting phases, for the acoustic beam transmission on the Mark 50 torpedo so that the beam pattern is flat over a wide angle and has maximum power [108]. Because of the lengthy computation, we showed in Ref. [108] that by producing a training set using the Gerchberg algorithm to train a neural network with our fast split inversion algorithm [82], we can find the phase thousands of times faster by means of the neural network [83].

### 4.4.2 Inverse Problem for Diffractive Optical Elements

Synthesizing a diffractive optical element in an inverse diffraction problem involves starting with a desired intensity at the observation plane as shown in Figure 4.12 and

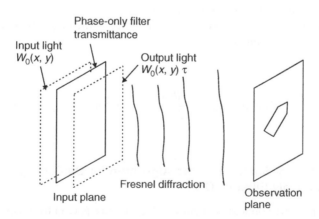

**FIGURE 4.12** Forward propagation from a diffractive optical element to an observation plane.

iteratively update the phases for the phase-only filter at the input plane. Mathematically, this involves solving the inverse nonlinear Fresnel diffraction integral equation (4.21) or Fraunhofer equation, in which nonlinearity is due to taking the modulus. That is, we solve for the phase-only input $U(x_1, y_1) = \exp\{j\phi_k\}$, given the output $|F_0(x, y)|$.

Note that because of the different nature of the nonlinearity here, this problem differs from the frequently arising conventional image processing problem of reconstructing an image from a distorted one in which a linear convolution-type problem with additive noise must be solved.

Furthermore, because phase wraps around at $2\pi$, there is significant ambiguity in the solution process. We did not consider phase unwrapping [46] because iterative algorithms converge to local minima and uniqueness of solution is not required. A typical goal for minimization is the root mean square (RMS) deviation of the modulus of the output from the desired output.

### 4.4.3   Gerchberg–Saxton Algorithm for Forward Computation

For the Fresnel forward computation (Figure 4.12), from the Fresnel diffraction equation (4.21), the magnitude of the output field $|F(x_0, y_0)|$ involves the magnitude of the convolution between the input field $U(x_1, y_1)$ and the Fresnel diffraction impulse response $h(x, y) = ((jk)/(2z)) \left[(x_0 - x_1)^2 + (y_0 - y_1)^2\right]$:

$$|F(x_0, y_0)| = \left| \frac{1}{j\lambda z} \int \int_{-\infty}^{\infty} U(x_1, y_1) \frac{jk}{2z} \left[(x_0 - x_1)^2 + (y_0 - y_1)^2\right] dx_1 \, dy_1 \right|$$

(4.21)

The observed intensity is $|F(x_0, y_0)|^2$.

We write the filter phase-only transmittance function as $\tau = \exp\{i\phi(x, y)\}$ and the input light field as $W_0$. The field after the input light has passed through the filter is then $U(x_1, y_1) = W_0\tau = W_0 \exp\{i\phi_0(x, y)\}$. Hence, the initial field at the input plane at the top left in Figure 4.13 is $W_0 \exp\{i\phi_0(x, y)\}$.

From equation (4.21), the expression for modulus of the output field at the observation plane can be written as

$$|F(x_0, y_0|$$
$$= \left| \int \int_{-\infty}^{\infty} W_0 \exp\{i\phi_0(x, y)\} \frac{jk}{2z} \left[(x_0 - x_1)^2 + (y_0 - y_1)^2\right] \right| dx_1 \, dy_1$$

(4.22)

### 4.4.4   Gerchberg–Saxton Inverse Algorithm for Designing a
### Phase-Only Filter or DOE

Designing a diffractive optical element involves finding $\phi(x, y)$ from equation (4.22) when the desired intensity of the observation plane image is defined as $I(y_0, y_0) = |F(x_0, y_0)|^2$.

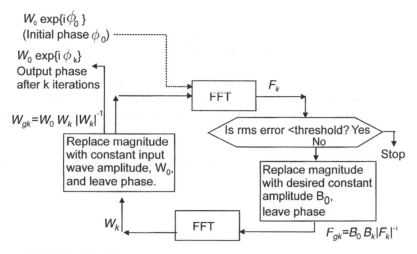

**FIGURE 4.13**  Diagram of Gerchberg–Saxton algorithm for DOE design.

The Gerchberg–Saxton algorithm uses an iterative successive approximation algorithm for solving this nonlinear integral equation, nonlinear because of the modulus sign in equation (4.22). First, we note that in practice the distance from input plane to observation plane is large compared to the transverse dimensions at the wavelength of interest. Therefore, as is known from diffraction theory [19], for large distances we can use the Fraunhofer approximation (Section 3.3.5), which allows us to use a single Fourier transform in place of the convolution with the Fresnel approximation (Section 3.3.4). The fast Fourier transform (FFT) is used to speed up the computation [133]. Figure 4.13 shows a diagram of the algorithm.

Prepare the initial input to start the iterative algorithm by choosing an initial phase $\phi_0$ (can be random) and compute $W_0 \exp\{i\phi_0(x, y)\}$ for iteration $k = 0$. Steps for the $k$th repeated loop are as follows:

1. Transform $W_k \exp\{i\phi_k(x, y)\}$ with an FFT to obtain $F_k(x_0, y_0)$.
2. Compute the error function $\epsilon_k$, the mean square error between the desired output $B(x_0, y_0)$ (which is real) and the amplitude of the output obtained at this iteration $|F_k(x_0, y_0)|$,

$$\epsilon_k = \int \int_{-\infty}^{\infty} \left[|F_k(x_0, y_0)| - B(x_0, y_0)\right]^2 dx_0 \, dy_0 \qquad (4.23)$$

Check if error function is sufficiently small to stop the iterations. Otherwise, repeat the loop.

3. Remove the magnitude of $F_k(x_0, y_0)$ by computing $F_k(x_0, y_0)|F_k(x_0, y_0)|^{-1}$ and replace the magnitude with that for the desired function, $B(x_0, y_0)$, that is,

compute

$$F_{gk}(x_0, y_0) = B(x_0, y_0)F_k(x_0, y_0)|F_k(x_0, y_0)|^{-1} \qquad (4.24)$$

4. Inverse Fourier transform $F_{gk}(x_0, y_0)$ to get $W_k(x_0, y_0)$.
5. Remove the magnitude of $W_k(x_0, y_0)$ by computing $W_k(x_0, y_0)|W_k(x_0, y_0)|^{-1}$ and replace the magnitude with that for the amplitude of the input wave, $W_0$, that is, compute

$$W_{gk}(x_0, y_0) = W_0 W_k(x_0, y_0)|W_k(x_0, y_0)|^{-1} \qquad (4.25)$$

# CHAPTER 5

# PROPAGATION AND COMPENSATION FOR ATMOSPHERIC TURBULENCE

Motion of air in the atmosphere is represented by active random fluctuations in velocity $V(x, y, z)$. The random velocity fluctuations are accompanied by random variations in temperature $(x, y, z)$. Expansion and contraction of air with temperature causes the density of the air to vary randomly, leading to passive random variations in refractive index $n(x, y, z)$. Here, passive means that variations in refractive index $n$ do not change the velocity of the air. Fluctuating changes in refractive index bend a beam of light randomly as it propagates through the atmosphere. Therefore, the beam wanders around and spreads out more than that for diffraction alone. When considering the effects of index of refraction on light beams, we refer to the atmospheric turbulence as optic turbulence.

There are two distinct situations of interest to the military: the effect on a coherent laser beam and the effect on incoherent imaging. The effect of turbulence on imaging is observable by the shimmering of a scene viewed across a tarmac parking lot that has been in the sun all day or by observing the twinkle of a star due to turbulence splitting the light into one or more paths that interfere, first explained by Isaac Newton. We will focus on the effects of spatially varying refractive index $n(x, y, z)$ due to turbulence on propagating laser beams. The resulting beams can be described by random fields $U(x, y, z)$ (or possibly $U(x, y, z, ct)$). Random fields vary in space and are the space equivalent of random processes [135], which vary only in time. *Stochastic* is often

*Military Laser Technology for Defense: Technology for Revolutionizing 21st Century Warfare*,
First Edition. By Alastair D. McAulay.
© 2011 John Wiley & Sons, Inc. Published 2011 by John Wiley & Sons, Inc.

used in place of *random*. Note that random fields are more complex than random processes because space has three dimensions while time has only one.

In Section 5.1, we review the locally homogeneous random field structure function and its power spectrum, following Ref. [4]. Those not needing to follow the detailed equations can skip this section. In Section 5.2, we describe optical turbulence: turbulence of refractive index that influences optical beams. Kolmogorov's cascade theory provides physical insight. Those not needing to follow detailed equations can skip over to Section 5.2.2 and equation (5.32). In Section 5.3, we describe adaptive optics devices and systems for cleaning up power beams and for compensating for optical turbulence, used in the airborne laser (Section 12.2) and homing missile protection (Section 12.3.3). In Section 5.4, we describe computational methods for modeling turbulence and our research on computation of phase screens widely used for layers of a layered model. Such phase screens are used in the airborne laser testbed (Section 12.2.7.2). In particular, we studied and wrote the first paper that compared two approaches, frequency and spatial, in modeling turbulence for phase screens [90]. The first method takes a spatial Fourier transform (FT) of the Kolmogorov spectrum [88] and the second one computes a covariance matrix from the Kolmogorov turbulence structure function. Both methods were shown to satisfy Kolmogorov's spectrum, but they handle problems of uncertain large atmospheric eddies differently and have different capabilities for modeling wind.

## 5.1  STATISTICS INVOLVED

Owing to random fluctuations of refractive index in space, whenever a pulse of light from a laser propagates through the atmosphere, it takes a slightly different physical path. The collection, called ensemble, of different paths in space composes a continuous random field [4]. Each path is considered to be a realization of the random field in space. A random *field* for which realizations are in space contrasts a random *process* for which realizations are in time. Because of randomness, the path for a realization is unpredictable, so we use averages or expected values that we label with $\langle a \rangle$ (in some texts E{} is used). For tractability, we initially focus on first- and second-order statistics: the mean and autocovariance. The mean for a random field $m(\mathbf{R})$ is the ensemble average at a point in space identified by its vector location $\mathbf{R}(x, y, z)$ over all realizations or paths:

$$m(\mathbf{R}) = \langle u(\mathbf{R}) \rangle \tag{5.1}$$

The spatial autocovariance function (second-order statistic) provides the expected relationship between the fields at two different locations $\mathbf{R}_1$ and $\mathbf{R}_2$:

$$B_u(\mathbf{R}_1, \mathbf{R}_2) = \langle [u(\mathbf{R}_1) - m(\mathbf{R}_1)][u^*(\mathbf{R}_2) - m^*(\mathbf{R}_2)] \rangle \tag{5.2}$$

When the mean is zero, $m(\mathbf{R}) = 0$, $B_u$ reduces to the autocorrelation function, a special case of the autocovariance function.

If $B_u$ does not depend on the absolute value of $\mathbf{R}(x, y, z)$ but only on the difference $\mathbf{R} = \mathbf{R}_2 - \mathbf{R}_1$ or $\mathbf{R}_2 = \mathbf{R}_1 + \mathbf{R}$, we have a statistically homogeneous field in space. (This is the equivalent of a wide sense stationary process in random processes.)

$$B_u(\mathbf{R}_1, \mathbf{R}_2) = B_u(\mathbf{R}_1 - \mathbf{R}_2) = \langle u(\mathbf{R}_1)u^*(\mathbf{R}_1 + \mathbf{R}) \rangle - |m|^2 \qquad (5.3)$$

Often fields are invariant under rotation or statistically isotropic: independent of direction of $\mathbf{R}$, $R = |\mathbf{R}_2 - \mathbf{R}_1|$. This allows us to reduce the 3D vector location to a 1D scalar, the distance $R$ from the axis of a beam. Then,

$$B_u(\mathbf{R}_1, \mathbf{R}_2) = B_u(R) \qquad (5.4)$$

### 5.1.1  Ergodicity

In practice, when we shine a laser beam through the atmosphere, we have only a single realization. Statistics for this one beam are easier to compute than those for an ensemble of beams. For example, a 3D space average is

$$\overline{B_u(\mathbf{R})} = \lim_{x \to \infty} \frac{1}{L} \int_0^L u(\mathbf{R})\mathrm{d}\mathbf{R} \qquad (5.5)$$

where $L$ is the path length. This is the 3D space equivalent of a time average in a random process

Similarly, for the space average for autocovariance,

$$\overline{B_u(\mathbf{R}_1, \mathbf{R}_2)} = \lim_{x \to \infty} \frac{1}{L} \int_0^L [u(\mathbf{R}_1) - m(\mathbf{R}_1)][u^*(\mathbf{R}_2) - m^*(\mathbf{R}_2)]\mathrm{d}\mathbf{R} \qquad (5.6)$$

Often we assume that the random field is ergodic, which means that the ensemble average at a point in space for a specific realization or path can be replaced by the space average for a single path. Ergodicity is difficult to prove except for simple cases, but for optical turbulence most of the time, ergodicity turns out to be a reasonable assumption. For an ergodic random field, the space average $\overline{u(\mathbf{R})}$ is the same as the ensemble average $\langle u(\mathbf{R}) \rangle$:

$$\overline{u(\mathbf{R})} = \langle u(\mathbf{R}) \rangle \qquad (5.7)$$

and the space average of autocovariance $\overline{B_u(\mathbf{R}_1, \mathbf{R}_2)}$ is the same as the ensemble average for the autocovariance $\langle B_u(\mathbf{R}_1, \mathbf{R}_2) \rangle$:

$$\overline{B_u(\mathbf{R}_1, \mathbf{R}_2)} = \langle B_u(\mathbf{R}_1, \mathbf{R}_2) \rangle \qquad (5.8)$$

Note that the space average for one realization is easier to measure than the ensemble average.

### 5.1.2  Locally Homogeneous Random Field Structure Function

In practice, for atmospheric turbulence, the mean is not constant over all space: velocity, temperature, and refractive index vary because the wind blows harder in some places. However, the mean is constant in a local region. We write

$$u(\mathbf{R}) = m(\mathbf{R}) + u_1(\mathbf{R}) \tag{5.9}$$

where $m(\mathbf{R})$ is the local mean that depends on $\mathbf{R}$ and $u_1(\mathbf{R})$ represents the random fluctuations about the mean. Locally homogeneous random fields are characterized by structure functions [4]. Such behavior frequently arises in random processes where it is labeled a random process with stationary increments [135], for example, a Poisson random process. Stationary may be qualified by wide sense when only autocovariance and mean are involved.

The structure function that characterizes locally homogeneous atmospheric turbulence is written as [4]

$$D_u(\mathbf{R}_1, \mathbf{R}_2) = D_u(\mathbf{R}) = \langle [u_1(\mathbf{R}_1) - u_1(\mathbf{R}_1 + \mathbf{R})]^2 \rangle \tag{5.10}$$

where we set $u(\mathbf{R}_1) \approx u_1(\mathbf{R}_1)$. Expanding equation (5.10) for a locally homogeneous random field gives

$$\begin{aligned} D_u(\mathbf{R}) = \langle u_1(\mathbf{R}_1)u_1^*(\mathbf{R}_1) \rangle + \langle u_1(\mathbf{R}_1 + \mathbf{R})u_1^*(\mathbf{R}_1 + \mathbf{R}) \rangle \\ - 2\langle u_1(\mathbf{R}_1 + \mathbf{R})u_1(\mathbf{R}_1) \rangle \end{aligned} \tag{5.11}$$

If the random field is also isotropic, $R = |\mathbf{R}|$, $D_u(\mathbf{R})$ is related to $B_u(\mathbf{R})$

$$D_u(R) = 2[B_u(0) - B_u(R)] \tag{5.12}$$

### 5.1.3  Spatial Power Spectrum of Structure Function

For high-frequency microwaves and optics, we often measure power because sensors may not be fast enough to record phase information. Furthermore, for random processes or fields, the phase may be of no consequence. Also, extracting signals from noise is often accomplished by spectral filtering to remove noise frequencies. So we need to compute the power spectrum $\Phi_u(\kappa)$ from the structure function $D_u(R)$ for locally homogeneous isotropic fields. Relation between a space function and its spectrum is usually defined using Fourier transforms (Section 3.1.2). This requires that the space function be absolutely integrable [133]:

$$\int_{-\infty}^{\infty} u(\mathbf{R})\mathrm{d}(\mathbf{R}) < \infty \tag{5.13}$$

Unfortunately, equation (5.13) is not valid for a random field or process. The Fourier–Stieltjes (or Riemann–Stieltjes) transform solves this problem [4, 135] and produces the well-known Wiener–Khintchine theorem.

For wide sense (mean and covariance) stationary random processes (in time), the autocovariance $B_x(\tau)$ and the power spectral density (or power spectrum) $S_x(\omega)$ are a Fourier transform pair:

$$S_x(\omega) = \frac{1}{2\pi} \int_{-\infty}^{\infty} B_x(\tau) \exp\{-i\omega\tau\}d\tau = \frac{1}{\pi} \int_0^{\infty} B_x(\tau) \cos(\omega\tau)d\tau$$

$$B_x(\tau) = \int_{-\infty}^{\infty} S_x(\omega) \exp\{i\omega\tau\}d\omega = 2 \int_0^{\infty} S_x(\omega) \cos(\omega\tau)d\omega \tag{5.14}$$

where the cosine equations arise in the second equalities because power and autocovariance are real and even.

In space, for statistically homogeneous isotropic random fields, from equations (5.14), the autocovariance $B_u(R)$ and 1D spatial power spectrum $V_u(\kappa)$ are similarly related by

$$V_u(\kappa) = \frac{1}{2\pi} \int_{-\infty}^{\infty} B_u(R) \exp\{-i\kappa R\}dR = \frac{1}{\pi} \int_0^{\infty} B_u(R) \cos(\kappa R)dR$$

$$B_u(R) = \int_{-\infty}^{\infty} V_u(\kappa) \exp\{i\kappa R\}d\kappa = 2 \int_0^{\infty} V_u(\kappa) \cos(\kappa R)d\kappa \tag{5.15}$$

where $\kappa$ is the spatial angular frequency, $\kappa = 2\pi f_s$ (space equivalent to $\omega = 2\pi f$).

We now find the relationship between the 1D spatial power spectrum $V_u$ in equation (5.15) and the spatial power spectrum for a 3D statistically homogeneous isotropic medium (for which radius $R = |R_2 - R_1|$). Note that these are not the same. For a statistically homogeneous medium (not isotropic), the 3D autocovariance $B_u(\mathbf{R})$ and the 3D spatial power spectral density are related by the Wiener–Khintchine theorem:

$$\Phi_u(\kappa) = \left(\frac{1}{2\pi}\right)^3 \iiint_{-\infty}^{\infty} \exp\{-i\kappa \cdot \mathbf{R}\} B_u(\mathbf{R})d^3\mathbf{R}$$

$$B_u(\mathbf{R}) = \iiint_{-\infty}^{\infty} \exp\{i\kappa \cdot \mathbf{R}\} \Phi_u(\kappa)d^3\kappa \tag{5.16}$$

If the medium is also isotropic, we set $R = |R_2 - R_1|$ and $\kappa = |\kappa|$ in equations (5.16) to get [4]

$$\Phi_u(\kappa) = \frac{1}{2\pi^2\kappa} \int_0^{\infty} B_u(R) \sin(\kappa R)R\,dR$$

$$B_u(R) = \frac{4\pi}{R} \int_0^{\infty} \Phi_u(\kappa) \sin(\kappa R)\kappa\,d\kappa \tag{5.17}$$

Therefore, the 3D spatial power spectrum for a locally statistically homogeneous isotropic medium, $\Phi_u(\kappa)$ in equation (5.17), is related to the 1D spatial power spectrum for a locally statistically homogeneous medium, $V_u(\kappa)$ in equation (5.15), by

$$\Phi_u(\kappa) = -\frac{1}{2\pi\kappa}\frac{dV_u(\kappa)}{d\kappa} \tag{5.18}$$

From equation (5.18), we see the origin of the $-11/3$ power rule that arises in models of turbulence: If 1D spectrum $V_u \propto \kappa^{-5/3}$, then the 3D spectrum $\Phi_u \propto -(5/3)V^{-8/3}/\kappa = -(5/3)V^{-11/3}$. Note that $\Phi_u(\kappa)$ in equation (5.17) has a singularity at $\kappa = 0$, so the convergence of $B_u(R)$ in equation (5.17) require restrictions in the rate of increase of $\Phi_u(\kappa)$ as $\kappa \to 0$ [4].

In practice, we allow a beam to propagate from one plane to another at separation $z$ so that we need the 2D spatial power spectrum $F_u(\kappa_x, \kappa_y, 0, z)$ between two planes $z$ apart [4]:

$$F_u(\kappa_x, \kappa_y, 0, z) = \int_{-\infty}^{\infty} \Phi_u(\kappa_x, \kappa_y, \kappa_z)\cos(\kappa_z z)d\kappa_z$$

$$\Phi_u(\kappa_x, \kappa_y, \kappa_z) = \frac{1}{2\pi}\int_{-\infty}^{\infty} F_u(\kappa_x, \kappa_y, 0, z)\cos(\kappa_z z)d\kappa_z \tag{5.19}$$

The structure function for a 3D locally homogeneous random field, $D_u(R)$, equation (5.11), is related to its 3D spatial power spectrum $\Phi_u(\kappa)$, equation (5.16), by [4, 60, 155]

$$D_u(\mathbf{R}) = 2\iint_{-\infty}^{\infty} \Phi_u(\kappa)[1 - \cos(\kappa \cdot \mathbf{R})]d^3\kappa \tag{5.20}$$

If the field is also isotropic, we have the Wiener–Khintchine equation pair

$$\Phi_u(\kappa) = \frac{1}{4\pi^2\kappa^2}\int_0^{\infty}\frac{\sin(\kappa R)}{\kappa R}\frac{d}{dR}\left[R^2\frac{d}{dR}D_u(R)\right]dR$$

$$D_u(R) = 8\pi\int_0^{\infty}\kappa^2\Phi_u(\kappa)\left(1 - \frac{\sin(\kappa R)}{\kappa R}\right)d\kappa \tag{5.21}$$

We note that some approaches to optical turbulence use the spatial power spectrum and some the structure function. We compare the two approaches in Refs. [88, 90].

## 5.2 OPTICAL TURBULENCE IN THE ATMOSPHERE

Turbulence in the atmosphere, or elsewhere, is considered to be due to statistical variations in velocity flow in a medium. The active variations in velocity cause pressure and temperature variations. Air expansion and contraction with temperature produces passive refractive index fluctuations: passive means these do not affect velocity. In a stream or a culvert, when there is no velocity mixing, the flow is laminar. A protrusion into the stream creates velocity mixing that generates turbulent eddies. Turbulence occurs when the Reynold's number, $Re$, exceeds

$$Re = \frac{Vl}{\nu} \tag{5.22}$$

where $V$ is the velocity, $l$ is the dimension, and $\nu$ is the viscosity in $m^2/s$. The Navier–Stokes partial differential equations for turbulence are generally too computationally demanding. So Kolmogorov developed a statistical approach [68] that is a mix of statistics and intuition, and is widely used today.

### 5.2.1  Kolmogorov's Energy Cascade Theory

When the velocity exceeds the Reynold's number, large size eddies appear of approximate dimension $L_0$ (Figure 5.1). Inertial forces break the eddies to decreasing sizes until a size $l_0$ is reached at which the absorption rate is greater than the energy injection rate and the eddies disappear through dissipation. Hence, at any time, there are a discrete number of eddies ranging in size from $l_0$ to $L_0$, the inertial range. Like whirlpools these eddies are somewhat circular and the air increases in velocity with radius and hence decreases in density with velocity. Lower density implies lower dielectric constant, so these eddies act like little convex lenses fluctuating around a mean orientation and of various sizes. Intuitively, such lenses cause a beam to wander and to spread more than that due to diffraction alone. From the cascade theory, structure functions and corresponding spatial power spectra can be developed for velocity, temperature, and refractive index; the latter results in optical turbulence.

The structure function for velocity may be written as

$$D_{RR}(R) = \langle (V_2 - V_1)^2 \rangle = C_v^2 R^{2/3} \quad \text{for} \quad l_0 \ll R \ll L_0 \tag{5.23}$$

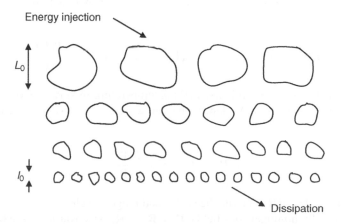

**FIGURE 5.1**  Energy cascade theory of turbulence.

where $C_v^2 = 2\xi^{2/3}$, the velocity structure constant $C_v$, is a measure of the total energy and $\xi$ is the average energy dissipation rate. For small-scale eddies, the smallest size $l_0 \approx \eta = v^3/\xi$, so the minimum size is less for strong turbulence. In contrast, the largest size is larger for strong turbulence because $L_0 \propto \xi^{1/2}$.

From the equations in Section 5.1, the 3D spatial power spectrum for a locally homogeneous isotropic random field can be written as

$$\Phi_{RR}(\kappa) = 0.033 C_v^2 \kappa^{-11/3} \quad \text{for} \quad l/L_0 \ll \kappa \ll 1/l_0 \tag{5.24}$$

Note that in equation (5.24), the 2/3 power law in the structure function for the locally homogeneous isotropic random field becomes a 5/3 power law for the 1D spectrum that becomes a characteristic (of Kolmogorov models) $-11/3$ power law in a 3D isotropic spectrum, as discussed in the text following equation (5.18).

Similarly, a spectrum may be written for temperature fluctuations as

$$\Phi_T(\kappa) = 0.033 C_T^2 \kappa^{-11/3} \quad \text{for} \quad l/L_0 \ll \kappa \ll 1/l_0 \tag{5.25}$$

where $C_T$ is the temperature structure constant.

Optical turbulence is associated with refractive index written by separating out the mean $n_0 = \langle n(\mathbf{R}, t) \rangle$:

$$n(\mathbf{R}, t) = n_0 + n_1(\mathbf{R}, t) \tag{5.26}$$

Normally, the variations of refractive index due to turbulence are much slower than the time taken for a beam to travel through it at the speed of light, so we can neglect time to obtain

$$n(\mathbf{R}) = n_0 + n_1(\mathbf{R}) \tag{5.27}$$

In the optical (including IR) wave bands, an empirical equation for refractive index is [4]

$$n(\mathbf{R}) = 1 + 77.6 + 10^{-6}(1 + 7.52 \times 10^{-3}\lambda^{-2})\frac{P(\mathbf{R})}{T(\mathbf{R})} \tag{5.28}$$

where $P(\mathbf{R})$ is pressure in millibars, $T(\mathbf{R})$ is temperature in kelvin, and $\lambda$ is optical wavelength in micrometers. As dependence on wavelength is weak, we can let $\lambda = 0.5\,\mu\text{m}$ to obtain

$$n(\mathbf{R}) = 1 + 79 \times \frac{P(\mathbf{R})}{T(\mathbf{R})} \tag{5.29}$$

The refractive index $n(\mathbf{R})$ inherits the scales and range of velocity.

For statistically homogeneous fields, $\mathbf{R} = \mathbf{R}_2 - \mathbf{R}_1$, so covariance $B_n(\mathbf{R}_1, \mathbf{R}_2) = \mathbf{B_n}(\mathbf{R}_1, \mathbf{R}_1 + \mathbf{R}) = \langle n_1(\mathbf{R}_1)n_2(\mathbf{R}_1 + \mathbf{R}) \rangle$ (Section 5.1). If the field is also isotropic,

$R = |\mathbf{R}_1 - \mathbf{R}_2|$, the structure function for refractive index can be written by analogy with equation (5.12) as

$$D_n(R) = 2[B_n(0) - B_n(R)] = \begin{cases} C_n^2 R_0^{-4/3} R^2 & 0 \le R \ll l_0 \\ C_n^2 R^{2/3} & l_0 \ll R \ll L_0 \end{cases} \tag{5.30}$$

where the strength of the turbulence is described by the refractive index structure parameter $C_n^2$ in units of $m^{-2/3}$. The inner scale is $l_0 = 7.4\eta = 7.4(\nu^3/\xi)^{1/4}$. $C_T^2$ in equation (5.25) may be obtained by point measurements of the mean square temperature difference with fine wire thermometers. $C_n^2$ can be computed from $C_T^2$ as

$$C_n^2 = \left(79 \times 10^{-6} \frac{P(\mathbf{R})}{T(\mathbf{R}^2)}\right)^2 C_T^2 \tag{5.31}$$

From Ref. [4], values of $C_n$, measured over 150 m at 1.5 m above the ground, range from $10^{-13}$ for strong turbulence to $10^{-17}$ for weak turbulence.

### 5.2.2 Power Spectrum Models for Refractive Index in Optical Turbulence

The following models enable generation of synthetic turbulence that may be used to estimate the effects of turbulence on optical propagation and allow test with adaptive optics.

- *Power Spectrum for Kolmogorov Spectrum*: From Section 5.2.1, the Kolmogorov spectrum for refractive index $n$ has the same form as that for velocity; therefore, replacing $C_v$ in equation (5.24) with $C_n$ gives

$$\Phi_n(\kappa) = 0.033 C_n^2 \kappa^{-11/3} \quad \text{for} \quad 1/L_0 \ll \kappa \ll 1/l_0 \tag{5.32}$$

  To compute phase screens to represent the effects of turbulence in space, we need to take an inverse Fourier transform of the spectrum. So we would like the model to have a range $0 \ll \kappa \ll \infty$. Unfortunately, extending the lower (large-scale) limit in equation (5.32) to zero wave number or spatial frequency is not possible because $\Phi_n$ blows up to $\infty$ as $\kappa \to 0$ because of the singularity at $\kappa = 0$, and this makes the evaluation of the Fourier integral impossible. Truncating the spectrum at the large-scale limit where frequency is lowest is undesirable because equation (5.32) shows that the energy in the turbulence is largest at this limit, leading to large abrupt transitions that provide an unrealistic spectrum.
- *Tatarski Spectrum and Modified Von Karman Spectra*: The Kolmogorov spectrum can be modified to extend above and below the limits by suitable multipliers to equation (5.32). For extension into the high wave numbers (spatial frequencies) past the small-scale limit $\kappa = 1/l_0$, caused by dissipation, we add

a Gaussian exponential proposed by Tatarski [155]. For extending into the lower wave number (low spatial frequency) past the large-scale limit $\kappa = 1/L_0$, we divide by a factor $(\kappa^2 + \kappa_0^2)$, where $\kappa_0 = C_0/L_0$ with $C_0 = 2\pi, 4\pi$, or $8\pi$ depending on the application [4]. The reason for the variety in selecting $C_0$ is that above the large-scale limit $L_0$, the field is no longer statistically homogeneous—the eddies may be distorted by atmospheric striations—so approximation is mainly to make the equations more tractable. The resulting modified Von Karman spectrum is

$$\Phi_n(\kappa) = \frac{0.033C_n^2}{(\kappa^2 + \kappa_0^2)^{11/6}} \exp\left\{\frac{-\kappa^2}{\kappa_m^2}\right\} \quad \text{for} \quad \kappa_m = 5.92/l_0, \kappa_0 = 2\pi/L_0$$

(5.33)

### 5.2.3   Atmospheric Temporal Statistics

When wind blows across the path of a laser beam in the atmosphere, the turbulence will move at wind speed $\mathbf{V}_\perp$ across the beam. Because the eddy pattern takes many seconds to change, the motion of noticeable wind $L_0/\mathbf{V}_\perp$ is much faster. This is known as the frozen turbulence hypothesis and is similar to cloud patterns caused by air turbulence passing across the sky. Therefore, we can convert spatial turbulence due to eddy currents into temporal turbulence that incorporates the effects of the component of wind at right angles to the path.

$$u(\mathbf{R}, t + \tau) = u(\mathbf{R} - \mathbf{V}_\perp \tau, t)$$

(5.34)

### 5.2.4   Long-Distance Turbulence Models

For an optical beam traversing many miles, the strength of turbulence may vary over the path. For example, if a beam moves over the land past cliffs over sea, the turbulence strength will vary because the land and sea have different temperatures and wind striking the cliff will generate strong turbulence. Such situations may be handled by partitioning the path into regions, each with its own constant turbulence. Further in numerical computations, the effect of turbulence in each region may be equated to a phase screen. Such phase screens were implemented around a sequence of optical disks for a hybrid test facility for the airborne laser (Section 12.2.7.2).

## 5.3   ADAPTIVE OPTICS

Adaptive optics may be used to reduce the impact of turbulence and is used in the Airborne Laser program (Section 12.2).

### 5.3.1   Devices and Systems for Adaptive Optics

There are two principal devices: wave front sensor to measure the shape of the wave front and deformable mirror to change the wave front.

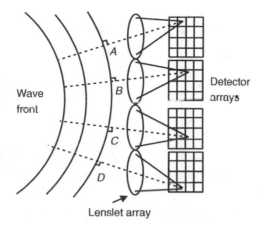

**FIGURE 5.2** Hartmann wave front sensor.

**5.3.1.1 Wave Front Sensor** The wave front sensor measures the shape of the wave front [144, 160, 163]. Many approaches are possible. Figure 5.2 illustrates how wave front shape is measured with a popular Hartmann wave front sensor. At $A$ in Figure 5.2, the wave front is tilted up substantially and the small lens will focus the light into the upper row of the associated sensor array. At $B$ the wave front is tilted up less, causing light to be focused to the next from top row of the sensor array. The lower half of the wave front focuses to the lower parts of the sensor array. The data from the wave front sensor arrays indicate the local direction in transverse $x$ and $y$ of the wave front.

The wave front shape is fed to a computer that calculates the settings for the deformable mirror to give a desired correction to the wave front. The computation allows the discretization of the wave front sensor and the deformable array.

**5.3.1.2 Deformable Mirror Device** The deformable mirror consists of a flexible mirror supported by an array of pistons that can raise the mirror up locally under computer control. Moving the mirror up locally advances the phase of the reflected light at this point because the light travels less far than that at neighboring points.

The light emanating from most high-power lasers has poor spatial coherence because of nonlinear effects that are significant at high powers (Section 8.1.2). Figure 5.3a shows the distorting effect of a beam of poor spatial coherence. Some parts of the wave front, labeled $A$, $B$, $C$, $D$, and $E$, are in the direction of the beam. But other points on the wave front point out of the beam along $F$, $G$, $H$, and $J$ and degrade the quality of the beam with distance. In a beam cleanup, the wave front is flattened into a plane wave by setting the deformable mirror to the conjugate phase of the wave front. For example, we can move to the right the mirror element of the deformable mirror that lines up with the wave front in the direction along the beam at $C$ (Figure 5.3a). This will retard the wave from $C$, so it has to travel farther than the one from $B$, which brings waves from $B$ and $C$ into line. Then waves propagating at an angle to the main beam ($F$, $G$, $H$, $J$) will be eliminated. In practice, adaptive optics

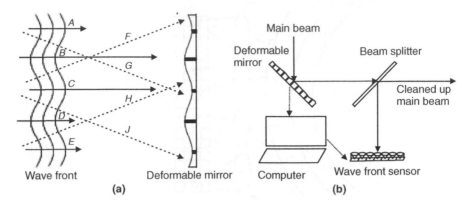

**FIGURE 5.3**   (a) Poor spatial coherence of a beam and (b) beam cleanup adaptive optics system.

will shape the wave front into a desired smooth function in space while improving beam spatial coherence.

### 5.3.1.3  *Adaptive Optics Cleanup System*   Figure 5.3b illustrates the principles of an adaptive optics beam cleanup system to improve spatial coherence and shape the wave front for a high-power laser. The adaptive optics beam cleanup system has an advantage over a spatial filter (Section 1.3.6), which cannot focus a poor quality beam (such as the one distorted by turbulence or from very high-power gas lasers) through a pinhole without losing considerable power. The spatial filter is adequate for cleaning up distortion caused by a Nd:YAG power amplifier (Section 13.2.1). The main beam to be cleaned up is deflected by a deformable mirror that will be set to perform the cleanup [83, 163]. The deformable mirror sends some of the reflected waves to the wave front sensor via a beam splitter so that the computer can determine how to set the deformable mirror to achieve beam cleanup and transform the wave front to that desired, as specified in the computer. Note that the beam size can be modified to operate with deformable mirrors of different sizes. For example, the $1000 \times 1000$ mirror array in the digital light projector (DLP) used in projectors, manufactured by Texas Instruments, is very small because it is a modified memory chip [83, 113]. The adaptive beam cleanup system (Figure 5.3), is incorporated into the airborne laser (ABL) system in Section 12.2.2.

### 5.3.1.4  *Principles of Adaptive Optics Compensation for Turbulence*
The main beam reflection from a missile cannot be relied on to provide adequate information for estimating turbulence. For this reason, a separate beacon laser is used that is coincident with the path of the main laser beam. The beacon laser light strikes the target and, unlike the main beam, is reflected back to the aircraft for the purpose of estimating the turbulence (Section 5.2). The beacon laser is a 1.06 μm diode pulsed semiconductor pumped solid-state laser (Section 8.2). Beacon lasers are used in adaptive optics for many astronomical telescopes. In this case, the beacon laser acts

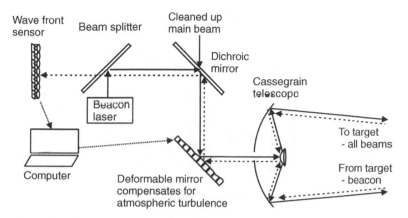

**FIGURE 5.4** Principle of adaptive optics for compensating for turbulence.

like the orange ranging beam that measures distance to set focusing before taking a photograph in many digital cameras.

The beacon laser is activated to allow compensating corrections to the wave front with a deformable mirror, *DMturb* in Figure 12.4, before shooting a pulse of the main high-power beam. Figure 5.4 shows the principle of using a beacon laser to estimate turbulence and then precompensate the main beam wave front. In Figure 5.4, the beacon laser light bounces off a dichroic mirror that reflects the Nd:YAG beacon beam (Section 8.2), at 1.06 μm and passes the COIL laser main beam (Section 8.3.2) at 1.315 μm. The beacon beam now bounces off the deformable mirror before being focused onto a spot on the target with a Cassegrain telescope (Section 1.3.4.1). The reflected beacon light returning from the target, distorted by two-way turbulence, bounces off the deformable and dichroic mirrors to pass through the beam splitter onto the wave front sensor. A computer iteratively calculates the setting for the deformable mirror to remove the effects of turbulence on the beacon beam. After the deformable mirror for removing turbulence stabilizes, light passing to the target and back again will be turned into a plane wave with turbulence effects removed. The main weapon beam, after cleaning up as described in Section 5.3.1.3, emits a powerful pulse that is merged into the same optical path as the beacon beam on passing through the dichroic mirror. Note that the pulsed beacon and main beams, although following the same path, occur at different times. The adaptive compensation system for turbulence (Figure 5.4) is incorporated into the ABL system in Section 12.2.3.

## 5.4 COMPUTATION OF LASER LIGHT THROUGH ATMOSPHERIC TURBULENCE

Temperature variations in the air over the earth generate convection currents that cause turbulence in the velocity of the air. Turbulence results in eddies of various sizes, typically from a few millimeters to an upper bound corresponding to tens to

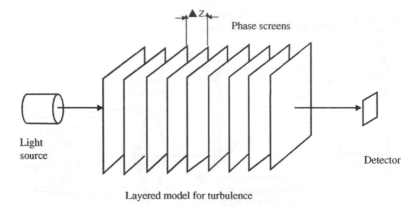

**FIGURE 5.5**  Layered propagation system for modeling propagation of light through turbulent atmosphere.

hundreds of meters. The latter corresponds to a low spatial frequency bound. The smallest eddy size decreases as the largest eddy size increases [155]. When small eddy size decreases, more samples in the transverse direction may be needed for accurate computation. The density variations of the air with velocity cause eddies in the refractive index that in turn bend and disperse the light. Turbulence affects the spatial and temporal properties of the laser beam and interacts with the diffraction process [4, 60, 144, 155].

Design and analysis of laser weapons (Section 12.2), optical communications [87], and optical imaging [144] require accurate simulation for the propagation of light through a turbulent atmosphere. Accurate modeling is required to develop and test adaptive optics systems (Section 5.3.1.4) necessary to compensate for turbulence with laser beams traveling long distances or near the earth. A substantial testbed facility was required for the Airborne Laser program (Section 12.2.7). It would be impossible to test and optimize such a system without a testbed with modeled turbulence because turbulence in the real world is extremely variable because of the variability due to weather conditions.

Computing turbulence with the Navier–Stokes equation is too computationally demanding, although polynomial models, fitted to Navier–Stokes results, have been attempted [84]. Therefore, statistical approaches and a layered model of the medium are typically used (Figure 5.5) [4, 88, 90]. The model developed here is used in the airborne laser testbed for adaptive optics and tracking in Section 12.2.7.

### 5.4.1  Layered Model of Propagation Through Turbulent Atmosphere

In the layered model (Figure 5.5), the propagation distance is split into layers of selected length. The atmospheric parameters are assumed to be constant in a layer and can differ between layers. Propagation across a layer is split into that due to diffraction in the absence of turbulence and that due to turbulence in the absence of diffraction. The two computations for each layer are performed in sequence. First,

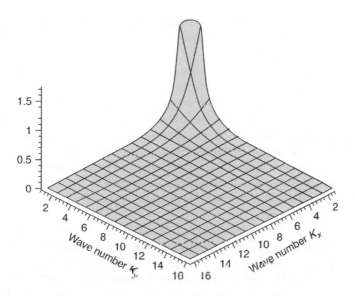

**FIGURE 5.6**   Kolmogorov spectrum in $x$ direction spatial frequency domain.

the light is propagated across a layer. Then, it is propagated through a phase screen that represents the effects of turbulence in traveling across the layer.

### 5.4.1.1   *Bounds on Spatial Frequency for Kolmogorov Power Spectrum Turbulence Model*   The Kolmogorov spectrum, as a function of spatial frequency in a transverse direction, is shown in Figure 5.6, equation (5.32). The Kolmogorov spectrum is specified only between limits specified by the curve shown. It is unspecified at spatial frequencies (or wavelengths) below or above this range. The lowest spatial frequencies correspond with the largest eddies determining the size of phase screen and the highest spatial frequencies correspond with the smallest eddies affecting the sampling fineness or resolution. Note that the power in the Kolmogorov spectrum increases with decreasing spatial frequency, so more energy arises at low spatial frequencies. Therefore, power below this lower bound can have a significant influence on the effect of turbulence. Unfortunately, the size of the largest eddies is not easily determined and is very much influenced by the weather and the application. As the power falls with decreasing eddy size, a lower resolution in space than that at the limit makes little difference.

### 5.4.1.2   *Application Dependence of Bounds for Kolmogorov Spectrum*
For beams interconnecting earth and satellites or high flying aircraft, the largest eddy size can reach a diameter of several hundred meters (Sections 12.2 and 12.2.7). Such large eddies may no longer be stationary for the beam propagation and some of the analysis may lose its validity.

For horizontal beams close to the earth, as for point to point optical wireless, the radius of the largest eddy size is limited by the distance of the beam from the earth.

A rule of thumb is sometimes used for which the largest eddy is approximately 0.4 times the height from the earth's surface [4]. In this case, there can be an abrupt cutoff for the lower spatial frequencies.

For optical wireless among sky scrapers in a large city, the largest turbulence eddy varies along the path, depending on the distance between buildings through which the beam passes and the height of buildings over which the beam passes. The fans in building air conditioners also generate eddies in the air. In the spectral method, as discussed in an earlier study [88], it was shown that an abrupt cutoff at low spatial frequencies presents a ringing problem in the Fourier transform process.

For laser beams angled up into the sky, the shape of the spatial spectrum is often not known and hard to measure because it may change with the weather conditions. However, the more gradual change in the largest eddy size makes the Fourier transform computation more accurate.

We next investigate the generation of turbulence phase screens for layers of the layer model (Figure 5.5) and compare two approaches: the first method is directly from the Kolmogorov spectrum and the second method is from the covariance of the structure function. The Kolmogorov approach starts with the Kolmogorov power spectrum and uses a Fourier transform to convert to a space screen, while the covariance approach starts in the space domain with the structure function for turbulence and computes the covariance. We show that in both methods the phase screens have similar appearance and, that on averaging, satisfy the Kolmogorov turbulence spectrum. However, they handle the limits in large eddies differently. We have limited ourselves to the case for which the phase screens in different layers are independent. In practice, if a wind is blowing approximately along the direction of beam propagation, the phase screens are related to a time delay and sideways velocity. The wind effects can be incorporated into the covariance method.

### 5.4.2  Generation of Kolmogorov Phase Screens by the Spectral Method

In this section, we generate phase screens directly from the Kolmogorov spectrum using methods described in the literature [23, 48, 77]. Phase screens are generated directly by Fourier transforming the Kolmogorov spectrum. In the Airborne Laser program, phase screens are written around the edge of several compact disks so that a laser beam in the laboratory will experience turbulence effects that are reproducible but can be changed by rotating the disks (Section 12.2.7.2, Figure 12.9).

#### 5.4.2.1  Equations for Generating Phase Screens by FT of Spectrum

The Kolmogorov spectral model for turbulence, equation (5.32), is [4, 144]

$$\Phi(f_x, f_y) = 0.033C_n^2(2\pi)^{-11/3}(f_x^2 + f_y^2)^{-11/6} \quad \text{for} \quad \frac{1}{L_0} \le f_x, f_y \le \frac{1}{l_0} \quad (5.35)$$

where $f_x = 2\pi/\kappa_x$ and $f_y = 2\pi/\kappa_y$ are the spatial frequencies in $x$ and $y$ transverse to the direction of propagation, $C_n^2$ is the strength of the turbulence, $\kappa$ is the transverse spatial angular frequency for the eddies, $L_0$ is the maximum eddy size, and $l_0$ is the minimum eddy size. Note that the spectrum is valid only in the range of $\kappa$ specified.

For propagation through a layer $n$ of thickness $\delta_z$ with light of wave number $k = 2\pi/\lambda$ and wavelength $\lambda$, the phase changes by [48]

$$(\Phi_n(f_x, f_y))^{1/2}$$
$$= (2\pi k^2 \delta_z)^{1/2} \left(\Phi(f_x, f_y)\right)^{1/2}$$
$$= 0.09844(\delta_z C_n^2)^{1/2}\lambda^{-1}(f_x^2 + f_y^2)^{-11/12} \quad \text{for} \quad \frac{1}{L_0} \le f_x, f_y \le \frac{1}{l_0} \quad (5.36)$$

The phase screen is the 2D Fourier transform of the left-hand side of equation (5.36). The right-hand side of this equation shows that this can be computed by taking the square root of the Fourier transform of the Kolmogorov spectrum, equation (5.35), and multiplying by a factor that depends on wavelength and layer thickness.

As discussed in Section 5.4.1, a problem arises if the spectrum is cut off abruptly below the lower bound as shown in Figure 5.6. The sharp transition, together with the cyclic nature of the discrete Fourier transform, results in slowly decaying ringing in the phase screen, which is an undesirable artifact of the spectral method.

Alternatives involving different approximations have been proposed for handling the low-frequency cutoff problem for direct transformation from the spectral domain. For example, the Von Karman spectrum (Section 5.2.2) [4, 144] gives a smooth function to zero at DC. Similarly, polynomials have been used to provide a smooth approximate for the spectrum [23]. For an airborne application such as the airborne laser, we chose to keep the power spectrum constant below the low spatial frequency cutoff.

### 5.4.2.2 Examples and Verification of Phase Screen Spectrum

Samples of random phase screens with a Kolmogorov spectrum are needed to represent turbulence in the simulator (Section 12.2.7.2). We follow a conventional approach [17] in which Gaussian white noise is multiplied by the desired spectrum and then the result inverse Fourier transformed. In this case, we create two 2D arrays of Gaussian random numbers for the real and imaginary parts of a complex random variable 2D array. The complex 2D array is multiplied by the square root of the Kolmogorov spectrum, equation (5.36). The sampling of the Kolmogorov spectrum in spatial frequency is selected in an interval of $\Delta f_x = 1/L_0$ so that the first sample is at the lower bound for the Kolmogorov spectrum. This sampling also provides adequate cross-section width for the widest anticipated beam. The real and imaginary parts of the Fourier transform provide two independent phase screens.

Each call to the phase screen generation function generates a 2D array of phases for a phase screen. As two phase screens are generated at one time only, every second call involves computation, unless the turbulence strength changes between the layers. Three sample phase screens are shown in Figure 5.7a–c. They vary due to the random model for turbulence. However, they conform to the same Kolmogorov spectral model. Therefore, they have the same spatial frequency content. As the Kolmogorov model has less high spatial frequencies, the phase screens have a similar spectrally smoothed

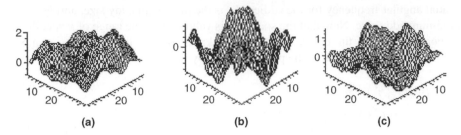

(a)                          (b)                          (c)

**FIGURE 5.7**   Sample phase screens using the *spectral* method for Kolmogorov turbulence model: (a) first, (b) second, and (c) third.

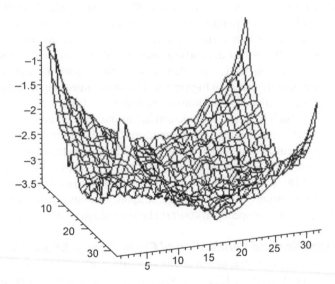

**FIGURE 5.8**   Average of absolute value of Fourier transform of 10 phase screens produced using the spectral method.

appearance. A noticeable hill in the phase screen acts as a lens on the light beam. As the hills are off-center, they will bend the light in a random manner, causing the beam wander at the receiver.

Correct implementation of the equations is verified by averaging the $|FT|^2$ for 100 random phase screens. The result is identical to that shown in Figure 5.6 and is shown in a 3D plot in Figure 5.8. One quadrant of Figure 5.8 matches Figure 5.6.

### 5.4.3    Generation of Kolmogorov Phase Screens from Covariance Using Structure Functions

We consider a much slower method that avoids taking a Fourier transform by computing a covariance matrix directly using a structure function [144]. The method is more versatile because it can include temporal dependencies between phase screens. A method for speeding the algorithm has been proposed [54]. However, the Kolmogorov spectrum is no longer truncated at the lower spatial frequency bound, resulting in excessive low spatial frequencies, unless a remedy is applied. We decided to truncate the low spatial frequencies, which represent very large-scale turbulence, by excluding the largest eigenvalues prior to computing the phase screens.

### 5.4.3.1 Equations for Computing the Phase Covariance for the Kolmogorov Spectrum

Structure functions are used to construct a covariance matrix from which phase screens are computed. The structure function for a homogeneous (spatially stationary) field $f$ with zero mean, as in equation (5.10), is [155]

$$D_f(r) = E\left[\{f(r_1 + r) - f(r_1)\}^2\right] \qquad (5.37)$$

$D_f(r)$ is large for periodic oscillations with wavelength less than $r$, providing the structure function with some spectral properties not present in a correlation function. Expanding equation (5.37) gives as in equation (5.12)

$$D_f(r) = 2[B_f(0) - B_f(r)] \qquad (5.38)$$

where $B_f(r)$ is the correlation function for $f$.

The equations for a phase covariance matrix are derived by assuming a phase across a transverse plane and subtracting the mean over the phase screen aperture. The covariance then has four correlation terms. The correlation terms of form $B_f(r)$ are replaced by structure functions $D_f(r)$ from equation (5.38). The $B_f(0)$ terms cancel. Note that this approach can be readily extended to the case in which phase screens at different layers are dependent [144]. The structure function for the Kolmogorov turbulence model is then used: (Similar to equation (5.23) for velocity)

$$D_n(r) = C_n^2 r^{2/3} \qquad (5.39)$$

The resulting expression for the phase covariance for one layer is

$$\gamma_\phi(n, m, n', m') \qquad (5.40)$$

$$= 6.88 r_0^{-5/3} \left\{ -\frac{1}{2} \left[ \{(m - m')\Delta x\}^2 + \{(n - n')\Delta y\}^2 \right]^{5/6} \right.$$

$$\left. + \frac{1}{2} \sum_{u',v'}^{N_x N_y} \frac{1}{N_x, N_y} \left[ \{(u' - m')\Delta x\}^2 + \{(v' - n')\Delta y\}^2 \right]^{5/6} \right.$$

$$+ \frac{1}{2} \sum_{u'',v''}^{N_x N_y} \frac{1}{N_x, N_y} \left[ \{(m - u'')\Delta x\}^2 + \{(n - v'')\Delta y\}^2 \right]^{5/6}$$

$$- \frac{1}{2} \sum_{u',v'}^{N_x N_y} \sum_{u'',v''}^{N_x, N_y} \left( \frac{1}{N_x, N_y} \right)^2 \left[ \{(u' - u'')\Delta x\}^2 + \{(v' - v'')\Delta y\}^2 \right]^{5/6} \Bigg\}$$

$$(5.41)$$

where the atmospheric coherence length or the Fried parameter is

$$r_0 = q \left[ \frac{4\pi^2}{k^2 C_n^2 \Delta z_i} \right]^{3/5} \tag{5.42}$$

The atmospheric coherence length $r_0$ is used to determine the size of aperture at the receiver and the maximum resolution when imaging through turbulence. The parameter $q$ is 0.185 for a plane wave and 3.69 for a spherical wave. The values for the Gaussian beam at the distances of interest are close to those for a plane wave. At $\Delta z = 1000$ m, $r_0 = 0.185$ m, which leads to an aperture of approximately 0.2 m$^2$. The propagation constant is $k = 2\pi/\lambda$ and the strength of the turbulence is taken as $C_n^2 = 10^{-13}$. There are many measurements and models for the strength of turbulence [4]. The distance along the propagation path for a layer $\Delta z_i \geq N_x \Delta x N_y \Delta y$. We used $N_x = N_y = 20$ and $\Delta x = \Delta y = 1$ in the simulator.

### 5.4.3.2 Computing a Kolmogorov-Based Phase Screen Sample for a Layer from the Covariance
The singular value decomposition (SVD) computes the eigenvalues, diagonals of $\Lambda$, and eigenvectors, columns of $U$, of the covariance matrix:

$$\Gamma = U \Lambda U^T \tag{5.43}$$

We generate a vector $\mathbf{b}'$ of uncorrelated Gaussian random variables with zero mean and variance $\Lambda$ of length $N_x \times N_y$. Then

$$E\left\{ \mathbf{b}' \mathbf{b}'^T \right\} = \Lambda \tag{5.44}$$

where $\Lambda$ is a matrix with eigenvalues on the diagonal and zero elsewhere. Random phase screens are formed using

$$\mathbf{a} = U \mathbf{b}' \tag{5.45}$$

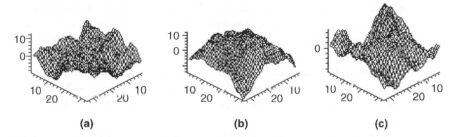

**FIGURE 5.9**   Sample phase screens using the *covariance* method for Kolmogorov turbulence model: (a) first, (b) second, and (c) third.

We verify that the new vector **a** has a covariance of $\Gamma$ by using equations (5.45) and (5.44):

$$E\left\{\mathbf{aa}^{T}\right\} E\left\{\mathbf{Ub'b'}^{T}\mathbf{U}^{T}\right\} \tag{5.46}$$

$$= \mathbf{U}E\left\{\mathbf{b'b'}^{T}\right\}\mathbf{U}^{T}$$

$$= \mathbf{U}\Lambda\mathbf{U}^{T} = \Gamma \tag{5.47}$$

### 5.4.3.3   *Examples of Covariance Method Phase Screens and Verification of Spectrum*   The computation of phase screens by the covariance method can be accelerated by noting that the covariance matrix can be scaled for different

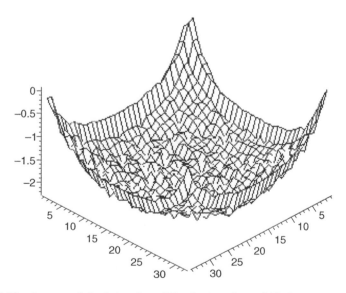

**FIGURE 5.10**   Average of absolute value of Fourier transform of 10 phase screens produced using the covariance method.

turbulence levels $C$, so that the eigenvalues and eigenvectors need to be computed only once at a cost of $O(p^2)$, where $p$ is the number of points in an $n \times n$ phase screen. Many phase screens can then be generated from one set of eigenvalues and eigenvectors by matrix multiplication, $O(p^2)$.

The covariance method does not have a mechanism for changing the low spatial frequencies away from the Kolmogorov spectrum equation. As a result, there are excessive low spatial frequencies. The effect is to have an overriding slope to the phase screen.

Most of the power in the Kolmogorov spectrum is at the low spatial frequencies. Similarly, the largest eigenvalues correspond to the largest power. Therefore, the low spatial frequencies can be reduced by decreasing the largest eigenvalues. The four largest eigenvalues for a $32 \times 32$ phase screen were set to zero to produce Figure 5.9.

Comparison of Figure 5.7a–c, produced with the spectral method, and Figure 5.9a–c, produced with the covariance method, shows similar looking phase screens. To verify that the covariance phase screens satisfy the Kolmogorov spectrum, we again averaged the absolute values for 10 screens, this time covariance generated ones (Figure 5.10). The result again agrees with the Kolmogorov spectrum shown in Figure 5.6, suggesting that the covariance matrix approach is as effective as the spectral method (Section 5.4.2) for generating phase screens. Although the covariance method requires slightly more computation than the spectral method, it allows wind effects to be included more easily.

# CHAPTER 6

# OPTICAL INTERFEROMETERS AND OSCILLATORS

Coherent light can be viewed as a wave or particle phenomenon. Waves can interfere with other waves because of their phase nature. If two waves with the same frequency and polarization are in phase, they will interfere constructively. That is, the resulting intensity, the average of the square of the sum of the two waves, will show increased intensity because of the wave overlap. In contrast, if the two waves are out of phase, the square of the sum of the waves will tend to zero intensity because of destructive interference. At phases between in-and out-of-phase, the average intensity will fall between that for in-phase and out-of-phase interference. Optical interference plays a critical role in military applications.

In Section 6.1, we describe the operation of interferometers, including three of the most common, Michelson, Mach–Zehnder, and Sagnac, for sensing a wide range of physical phenomena. In Section 6.2, we analyze the Fabry–Perot resonator, which traps light in a cavity between two parallel mirrors and resonates at frequencies for which reflection from the mirrors interferes constructively with waves trapped in the cavity. Fabry–Perot resonators are fundamental for many filters and lasers (Chapter 7) [83]. In Section 6.3, we consider multireflection interference filters made from thin layers of transparent dielectric materials of different media. Such layers are widely used for coating optical elements, for example, in eyeglasses, to avoid reflection. We describe the matrix method for design and analysis and an inverse method to measure optical properties of individual layers [95, 97].

*Military Laser Technology for Defense: Technology for Revolutionizing 21st Century Warfare*,
First Edition. By Alastair D. McAulay.
© 2011 John Wiley & Sons, Inc. Published 2011 by John Wiley & Sons, Inc.

## 6.1  OPTICAL INTERFEROMETERS

Interferometers provide sensing for measuring a variety of physical phenomena, such as distance, pressure, temperature, rotation (gyroscopes), presence of chemical or biological weapons, and perimeter protection sensors—all of which are important in military applications [16, 55]. An interferometer splits coherent light from a laser into two paths. In many applications, one path acts as a reference and the other experiences the phenomenon to be measured. The phenomenon to be measured changes the velocity of the light in the measurement path, causing the phase of light to differ from that in the reference path. Note that the output is influenced by both changes in absorption between the two arms and changes in refractive index. On recombining the light paths in the interferometer and using a photodetector or CCD camera, the phase difference is converted to an intensity difference indicative of the strength of the phenomena.

Coherent light from a laser has temporal and spatial coherence that affect the performance of an interferometer. Spatial coherence is a measure of how close a wave is to a plane wave. High spatial coherence is achieved with a small source (Sections 1.3.6 and 3.2.2). Temporal coherence concerns how close light is to a single frequency, monochromatic, and is discussed here.

The bandwidth in wavelength $\Delta\lambda$ of the source laser limits the maximum difference in distance, $l_c$, the coherence length, at which combining the two paths of an interferometer will produce a distinct interference pattern. At a difference in path lengths of $l_c$, the constructive interference from the center wavelength $\lambda$ is canceled by the destructive interference from the wavelength $\lambda + \Delta\lambda/2$ at the band edge. For constructive interference at a difference in distance of $l_c$, and a wavelength of $\lambda$, the two paths must differ by a whole number $m$ of wavelengths so as to be in phase. Therefore,

$$m\lambda = l_c \tag{6.1}$$

Destructive interference occurs at a whole number of wavelengths minus a half. Therefore, at a difference in distance of $l_c$ and a wavelength $\lambda + \Delta\lambda/2$ at the edge of the frequency band,

$$(m - 1/2)(\lambda + \frac{\Delta\lambda}{2}) = l_c \tag{6.2}$$

For the constructive interference at the nominal wavelength $\lambda$ to be canceled out by the destructive interference, we equate $l_c$ in equation (6.1) with $l_c$ in equation (6.2) and solve for integer $m$ to get

$$m = \frac{\lambda}{\Delta\lambda} + \frac{1}{2} \approx \frac{\lambda}{\Delta\lambda} \tag{6.3}$$

Substituting equation (6.3) into equation (6.1) gives a coherence length in terms of wavelength and bandwidth of the source:

$$l_c = \frac{\lambda^2}{\Delta\lambda} \tag{6.4}$$

Using equation (6.20) to convert from wavelength to a frequency source bandwidth $\Delta f$,

$$l_c = \frac{c}{\Delta f} \tag{6.5}$$

From equation (6.5), a corresponding coherence time $t_c = l_c/c$, the maximum time difference between two paths of an interferometer before the interference pattern disappears, is the reciprocal of the source frequency bandwidth (for a laser for which $\Delta f$ is small, the bandwidth is known as linewidth):

$$t_c = \frac{1}{\Delta f} \tag{6.6}$$

Interferometers may be implemented in free space on an optical bench. In the laboratory, optical tables with air supported legs are required to maintain stability. Often interferometers are implemented with optical fiber or in integrated optics [57].

In Section 6.1.1, we describe the Michelson interferometer, commonly used for accurate distance measurement among other things. In Section 6.1.2, we describe the Mach–Zehnder interferometer in free space, optical fiber, and integrated circuits. The Mach–Zehnder interferometer is used in integrated optics for sensing and on–off light switching, for example, for telecommunication modulators [102, 106]. In Section 6.1.3, we describe a stable optical fiber Sagnac interferometer, which is the basis of modern gyroscopes and novel perimeter security systems [112].

### 6.1.1   Michelson Interferometer

Figure 6.1 shows a beam splitter at the center that splits light into two equal parts: one passing to mirror $M_1$ and the other to mirror $M_2$. In this case, the Michelson interferometer measures the displacement of mirror $M_2$. Such interferometers are used in VLSI lithography machines and elsewhere for very accurate distance measurements.

Following reflection from the mirrors, the paths are combined at the screen after passing through the beam splitter again. Consequently, the field after combining the two paths at the screen may be written as

$$E(t) = E_1(t) + E_2(t) \tag{6.7}$$

for which intensity is

$$
\begin{aligned}
I = \langle EE^* \rangle &= \langle (E_1 + E_2)(E_1 + E_2)^* \rangle \\
&= \langle E_1 E_1^* \rangle + \langle E_2 E_2^* \rangle + \langle E_1 E_2^* \rangle + \langle E_2 E_1^* \rangle \\
&= I_1 + I_2 + 2\mathrm{Re}\left\{ \langle E_1^* E_2 \rangle \right\}
\end{aligned} \tag{6.8}
$$

If the intensities are equal for the fields, $I_1 = I_2$,

$$
I = 2I_1 + 2\mathrm{Re}\left\{ \langle E_1^* E_2 \rangle \right\} \tag{6.9}
$$

### 6.1.1.1 *Autocorrelation Function in Michelson interferometer*   For a Michelson interferometer, if mirror $M_2$ is moved to the right by distance $d$, as shown in Figure 6.1, the electric field $E_2$ will be delayed by time $\tau = 2d/c$ relative to field $E_1$, where $c$ is the speed of light in air. Therefore, as mirror $M_2$ is moved, $E_2(t) = E_1(t - \tau)$, and the Michelson interferometer computes the autocorrelation of the input beam. The product $E_1^* E_2$ in equation (6.8) can now be written with the autocorrelation function $\Gamma(\tau)$ for the Michelson interferometer. In optics, the auto-correlation function $\Gamma(\tau)$ is called the *complex self-coherence function*:

$$
\Gamma(\tau) = \langle E_1^*(t) E_1(t - \tau) \rangle \tag{6.10}
$$

So, substituting equation (6.10) into equation (6.9) allows the intensity on axis out of a Michelson interferometer to be written as

$$
I = 2I_1 + 2\mathrm{Re}\left\{ \Gamma(\tau) \right\} \tag{6.11}
$$

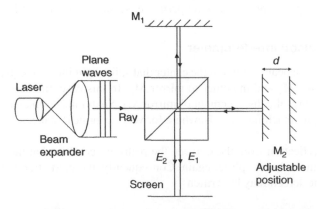

**FIGURE 6.1**   Michelson interferometer for measuring distance.

Assuming an ergodic process (Section 5.1.1), the autocorrelation function can be written as

$$\Gamma(\tau) = T_m \overset{\lim}{\to} \infty \frac{1}{T_m} \int_{-T_m/2}^{T_m/2} E_1^*(t)E_1(t-\tau)dt \tag{6.12}$$

Equations (6.10) and (6.12) indicate how well a function is correlated with itself delayed by time $\tau$. Note that a function well correlated with its shifts, with a broadband in $\tau$, allows prediction into the future—valuable for predicting stock prices. From the Wiener–Khintchine theorem, the Fourier transform of the autocorrelation function (Section 5.1.3) is the spectrum of the function, so such an easily predicted function has a narrow spectrum—in the limit, the future of a single-frequency sine wave in time is totally predictable.

For the case where light is sinusoidal in time, $E_1(t) = E_0 \exp\{-\omega t\}$, from equation (6.12),

$$\Gamma(\tau) = T_m \overset{\lim}{\to} \infty \frac{1}{T_m} \int_{-T_m/2}^{T_m/2} |E|_0^2(t) \exp\{i\omega t\} \exp\{-i\omega(t-\tau)\}dt$$

$$= |E_0|^2 \exp\{i\omega\tau\} = I_1 \exp\{i\omega\tau\} \tag{6.13}$$

From equations (6.11) and (6.13), the output of a Michelson interferometer can be written as

$$I(\tau) = 2I_1(1 + \cos \omega\tau) \tag{6.14}$$

Equation (6.14) is plotted in Figure 6.2. The peaks at $2\pi$ intervals in $\omega\tau$ are referred to as fringes. The spacing between peaks is equivalent to moving the mirror in Figure 6.1 by a distance $\lambda/2$, where $\lambda$ is the wavelength of light. Then for visible red light,

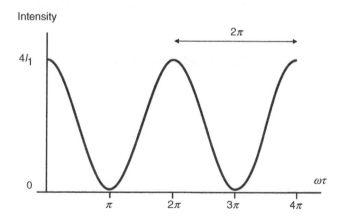

**FIGURE 6.2**   Output of Michelson interferometer.

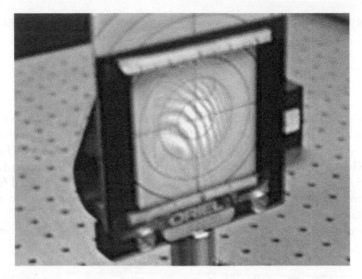

**FIGURE 6.3** Spatial interference pattern for Michelson interferometer.

$\lambda = 633$ nm, the mirror needs to be moved only a distance of $d = 633/2 = 317$ nm for a detector at the center of the Michelson output screen to vary from lowest to highest intensity, providing very accurate position measurement. This configuration is used for accurate distance measurement in VLSI lithography machines.

If we observe the whole screen, we see an interference pattern in space as shown in the photograph from the laboratory setup in Figure 6.3. An equation for the spatial pattern can be written similarly to that for the temporal pattern. The pattern arises because for different off-axis points on the output screen, the path lengths differ, causing interference to vary radially outward in space.

Note that if we use two different sources, the autocorrelation function or complex self-coherence function becomes a cross-correlation function, in optics a *cross-coherence function*,

$$\Gamma_c(\tau) = \langle E_1^*(t)E_2(t + \tau) \rangle \tag{6.15}$$

This may be normalized to

$$\gamma_c(\tau) = \frac{\Gamma_c(\tau)}{\Gamma_0} \tag{6.16}$$

The output of the Michelson interferometer can now be written as

$$I(\tau) = 2I_1 + 2\text{Re}\{\gamma(\tau)\} \tag{6.17}$$

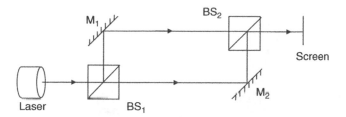

**FIGURE 6.4**   Mach–Zehnder free-space interferometer.


## 6.1.2   Mach–Zehnder Interferometer

The Mach–Zehnder interferometer may be implemented in free space, optical fiber, or integrated optics, depending on the application and manufacturing volume. The application is often first tested in free space but alignment and stability are detrimental for mass production. Optical fiber is used where distributed sensing is required. Integrated optics is used for point sensing when manufacturing volume is high because unit cost is reduced by volume and the initial design cost occurs only once.

***6.1.2.1   Free-Space Mach–Zehnder Interferometer***   A free-space setup is shown in Figure 6.4, in which beam splitter $BS_1$ is used to split light into two paths. The two paths are recombined by beam splitter $BS_2$. If the environment path is identical to the reference path, the light will be in phase at beam splitter $BS_2$ and constructively interfere to give back the original light intensity of the source at the center of the screen (minus reflective losses). If the environment changes the propagation time in one arm of the interferometer, there will be a phase difference at the screen. If phase difference is 180° ($\pi$), the beams at the center of the screen will combine destructively (destructive interference) and there will be no light at the center of the screen. Note that power is not destroyed—it just no longer illuminates the center of the screen.

A setup in our laboratory is shown in Figure 6.5. The two paths through the interferometer cannot differ by more than the coherence length (Section 6.1) of the laser, at which distance the fringes in time and space will vanish. In the case of the fiber optic and integrated optic versions following, the light is dispersed into the cladding on destructive interference.

***6.1.2.2   Optical Fiber Mach–Zehnder Interferometer***   Mach–Zehnder interferometers are often implemented in fiber [164]. Components may now be plugged into one another and optical alignment avoided, reducing assembly cost relative to free space. Figure 6.6 shows such a fiber Mach–Zehnder interferometer. A directional coupler has no back reflections and splits the light into two equal intensities along two fibers, one acting as reference and the other for sensing. When light crosses from one side to the other while propagating through a directional coupler, it experiences a 90° phase lag [176]. A second 50/50 coupler, acting as a combiner, merges the paths together. Therefore, the top path from input to output experiences no crossovers in the sensing path while the lower path has experienced two. The 180° phase difference

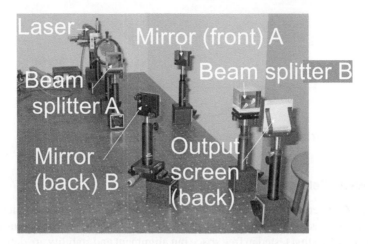

**FIGURE 6.5**    Photograph of Mach–Zehnder free-space interferometer in laboratory.

between top and lower paths results in zero output because of destructive interference. A photograph of a prototype of such a fiber interferometer is shown in Figure 6.7.

For maximum optical interference, light from the two paths must have the same frequency and polarization. Ordinary optical fiber does not maintain polarization, so a polarization controller (Section 2.2.5) is included in Figure 6.6. The optical fiber Mach–Zehnder interferometer often measures strain caused by stress in mechanical and structural systems. In columns and struts of bridges, strain due to excessive loads, such as tanks or heavily loaded trucks, or subsidence with age can be instantly spotted, and if excessive, a wireless signal is sent back to a highway monitoring office. In military applications, the Mach–Zehnder fiber interferometer may be used in the wings of military aircraft. Excessive strain to wings can be monitored to avoid exceeding damage limits while maneuvering abruptly. The strain sensor is particularly valuable if a wing is damaged and the pilot has to maneuver within a strain limit.

Optical sensors are advantageous in hazardous environments because there are no sparks when a fiber breaks or its cover is worn. This allows hot switching in optical telecommunication circuits. An aircraft on an international flight crashed a few years ago because of frayed wires inside the wing where fuel is stored. Optical fiber with

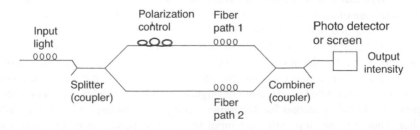

**FIGURE 6.6**    Optical fiber Mach–Zehnder interferometer.

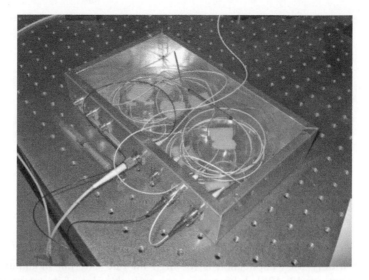

**FIGURE 6.7**  Laboratory prototype of an optical fiber Mach–Zehnder interferometer.

cladding removed in one arm makes an effective hazardous liquid fuel gage, in which the level of fuel is detected from the output of an interferometer.

Path length $p$ for a length of optical fiber with refractive index $n$ is the equivalent length traveled in air, $p = nL$. For many environmental sensors, the cover should be removed from the sensing arm. Care must be taken to measure only the parameter of interest. For example, in a temperature measuring device [57], the optical path length $p$ of the sensing side changes due to both thermal expansion $\alpha - dT/dL$ in °C per unit length and changes in refractive index with temperature, $dn/dT$. So the output depends on

$$\frac{dp(n, L)}{dT} = n\frac{dL}{dt} + L\frac{dn}{dT} \tag{6.18}$$

### 6.1.2.3  *Integrated Circuit Mach–Zehnder Interferometer*  The Mach–Zehnder interferometer is etched into a chip using lithographic techniques for integrated options [57] (Figure 6.8). The reference path must be protected from the environment to be measured.

Integrated optic Mach–Zehnder interferometers have been used for temperature sensors [57, 64] . If the environment is to be sensed, the optical field in the waveguide must not be tightly confined in the waveguide; that is, the exponentially decaying field from the surface wave [78] must penetrate the air above the integrated optic chip. In some cases, porous glass (sol-gel) is deposited over one arm of device. The pores can match the size of a chemical molecule or virus such as *Salmonella*. If a chemical is present, it accumulates by trapping in the pores, changing the refraction index or absorption of the light passing through that side of the waveguide.

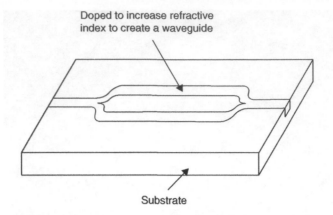

**FIGURE 6.8**   Integrated optic Mach–Zehnder interferometer.

Another widely used application is for high bit rate external modulations for telecommunications [2, 176]. In this case, one or both sides of its waveguide are filled with electro-optic material that changes refractive index with applied voltage. The output can be modulated between light and no light by applying an electric signal at up to 10 Gbps with communication information.

### 6.1.3   Optical Fiber Sagnac Interferometer

***6.1.3.1   Rotation Sensing***   The Sagnac interferometer is widely used for gyroscopes. All military mobile platforms such as airplanes, ships, and tanks use gyroscopes to orient themselves in three dimensions to maintain stability for their guns and weapons. Earlier, three orthogonal spinning tops were used to detect orientation and motion, but they are too slow for modern military. The mechanical gyros have been replaced with three-axis fiber gyroscopes based on Sagnac interferometers.

Figure 6.9a shows an optical fiber Sagnac interferometer for sensing rotation of a loop of optical fiber about its axis. In practice, several concentric loops are used to amplify the effect. Light from a laser source is split into two equal paths, one tracking clockwise (right circular) around the fiber loop and the other tracking anticlockwise (left circular) around the loop. As clockwise and anticlockwise paths pass through the

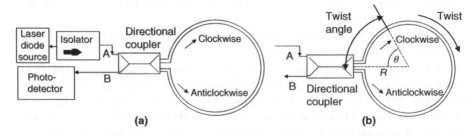

**FIGURE 6.9**   Sagnac interferometer: (a) at rest and (b) after rotation by $\phi$.

same fiber loop, light travels a meter in a nanosecond, and the environment usually takes milliseconds to change, both paths normally see the same environment in a static case. Consequently, the Sagnac interferometer is very stable, not needing to be balanced when the environment changes, unlike a Mach–Zehnder interferometer. In traveling from A to B (Figure 6.9) along the clockwise path, the light never crosses over to the other side in the coupler, whereas light traveling from A to B via the anticlockwise path crosses to the other side of the coupler twice. As discussed in Section 6.1.2.2, each crossing incurs a $\pi/2$ phase change. Therefore, in reaching B, the anticlockwise light lags the clockwise light by $\pi$ radians or 180°. Hence, they combine to zero by destructive interference at output B in Figure 6.9a. In other words, there is normally no output at B.

In contrast, both clockwise and anticlockwise paths experience a single crossover when traveling from A back to A. Thus, there is constructive interference on returning to A. An isolator blocks light from returning into the laser where it could interfere with or damage the laser.

In Figure 6.9b, the coil rotates clockwise $\Omega$ radians per second and is moved through an angle $\theta$ while light travels once around the loop. During this time, due to the rotation of the coil, the light traveling in the clockwise path travels farther to reach the coupler while that traveling in the anticlockwise direction travels less far. A phase difference due to rotation is observed at output B from which the rate of rotation $\Omega$ can be calculated. For a circular loop, the speed at which the coil moves is $\Omega R$ in the clockwise direction and $-\Omega R$ in the anticlockwise direction. The time for light to travel around the loop is approximately $2\pi R/(c/n)$, where $n$ is the refractive index of the fiber. Hence, the difference in distance traveled by the clockwise and counter-clockwise paths is (speed traveled by coil times time for light to travel around the loop) $d_f = 2\Omega R \times (2\pi Rn/c)$ and phase difference between paths is $\Delta\psi = kd_f = 2\pi n/\lambda \times 4\pi R^2 \Omega n/c = 8\pi^2 R^2 \Omega n^2/(\lambda c)$, where $k$ is the propagation constant in free space.

Students in my laboratory constructed and demonstrated a free-space optical gyroscope. A beam splitter directs light in two directions, clockwise and anticlockwise, in the same loop made up of three small mirrors at the corners of a square to form a Sagnac interferometer. Rotation of the system on a rotating platform (lazy Susan) results in light traveling different distances in the two directions to reach a common photodetector chip. The chip, used in vending machines, converts an intensity variation resulting from the rotation-generated phase difference into a rate of pulsing that directly activates an audio amplifier. Spinning the rotating platform results in the sound pitch increasing in proportion to the speed of rotation.

### 6.1.3.2 Perimeter Security

The Sagnac interferometer around a military facility or other vulnerable regions is used for perimeter security. Reference [112] describes our research into determining the location of a perimeter penetration.

## 6.2 FABRY–PEROT RESONATORS

Optical resonators are fundamental for generating laser light (Chapter 7) and for filtering [105] and amplification.

### 6.2.1 Fabry–Perot Principles and Equations

An optical resonator is formed by trapping light between two parallel vertical mirrors that form a cavity or an etalon; this type of resonator is known as a Fabry–Perot resonator [53, 176]. If we stand inside a pair of Fabry–Perot mirrors in a fairground hall of mirrors, we can see an infinite number of reflections. A different resonator, a ring resonator, is often used in integrated optics [96, 102, 105, 106]. In a Fabry–Perot resonator, photons bounce back and forth between the mirrors and at certain resonant wavelengths they coherently add inside the cavity to increase the power trapped in the resonator to a high level. This is like pushing a child's swing at the correct intervals for resonance to slowly increase the energy contained in the swing. The Fabry–Perot resonator can also enhance images and many fish and animals that live in dark environments, such as deep sea fish, have reflecting surfaces at the front and rear of their eyes to trap photons so that the intensity in the etalon is sufficient to excite neurons in the retina.

### 6.2.2 Fabry–Perot Equations

The equations for a Fabry–Perot resonator are derived by considering reflection and transmission at a single mirror. Partially reflecting mirrors allow input to and output from the cavity between them. Figure 6.10a and b shows for an incident field from the left the reflection and transmission at a mirror interface without attenuation and with power attenuation $A$, respectively. A negative power attenuation can be used to represent power gain. Note that the reflected power is $R$ times the input power. To preserve power, the output power must then be $1 - R$ times the input power. Hence, from Figure 6.10b, for an incoming field $E_i$, $\sqrt{R}$ is reflected and $\sqrt{1 - A - R}$ is transmitted by the mirror.

From Figure 6.10, we can hypothesize light bouncing back and forth between the mirrors, as shown in Figure 6.11. We assume a phase change in crossing the etalon of $\beta x$, where $\beta$ in radians per second is the propagation constant and $x$ is the etalon width. Hence, on reaching the right-hand side mirror in Figure 6.11, the field is $\sqrt{1 - R - A} E_i \exp\{j\beta z\}$ and on reflection from the mirror, $\sqrt{R}\sqrt{1 - R - A} E_i \exp\{j\beta z\}$. With perfect mirror alignment, the light reflects back and forth along the same path. For clarity in Figure 6.11, we shifted the beams down on reflection.

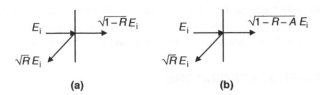

$$E_i \quad \xrightarrow{\quad} \quad \sqrt{1-R}\,E_i \qquad\qquad E_i \quad \xrightarrow{\quad} \quad \sqrt{1-R-A}\,E_i$$
$$\sqrt{R}\,E_i \swarrow \qquad\qquad\qquad \sqrt{R}\,E_i \swarrow$$

**(a)**                                              **(b)**

**FIGURE 6.10**  Reflection and transmission at a mirror: (a) without attenuation and (b) with attenuation.

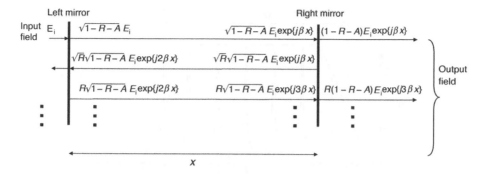

**FIGURE 6.11** Illustration for deriving Fabry–Perot transfer function equations.

From Figure 6.11, the output at the right of the etalon is the infinite sum of the outputs transmitted through the output mirror on each bounce of the light:

$$E_o = (1 - A - R)E_i \exp\{j\beta x\} \sum_{m=0}^{\infty} \left[ R \exp\{j2\beta x\} \right]^m \qquad (6.19)$$

where for each cycle of the light, we add in the sum $R$ for two reflections and $\exp\{j2\beta x\}$ for two propagations across the cavity. Similarly, we could determine the reflection from the resonator by summing the infinite reflections to the left.

If incoming broadband light has an electric field $E_i$ only those wavelengths will resonate for which a whole number of half-wavelengths fit between the mirrors (Figure 6.12).

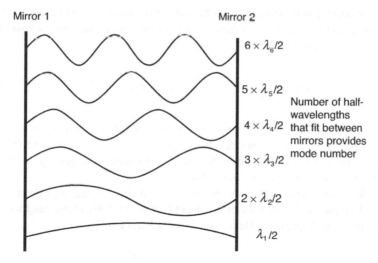

**FIGURE 6.12** Electric fields for modes in Fabry–Perot resonator.

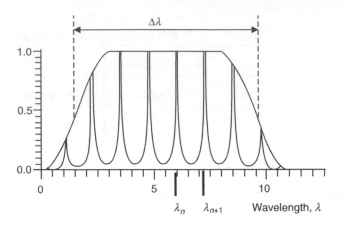

**FIGURE 6.13** Longitudinal modes in wavelength spectrum out of Fabry–Perot resonator.

Consequently, if the bandwidth of $E_i$ in wavelength is $\Delta\lambda$, as shown in Figure 6.13, there will be many resonance peaks corresponding to different numbers of half-cycles, $\lambda_n, \lambda_{n+1}, \ldots$, as shown in Figure 6.13.

Conversion between frequency $f$ and wavelength $\lambda$ and variations in frequency $\Delta f$ and wavelength $\Delta\lambda$ can be made from the respective equations,

$$f = \frac{c}{\lambda} \quad \text{and by differentiation} \quad |\Delta f| = \frac{c}{\lambda^2}\Delta\lambda \tag{6.20}$$

where air (refractive index $n = 1$) is assumed in the cavity.

### 6.2.2.1  Fabry–Perot Transfer Function

Propagation of a plane wave can be written in exponential form as $\exp j(\beta x - \omega\tau)$. So, in propagating once across etalon, the phase angle changes by

$$\beta x = -\frac{\omega}{c}x = -\omega\left(\frac{x}{c}\right) = -\omega\tau = -2\pi f\tau \tag{6.21}$$

where $x$ is the distance between mirrors across the air-filled etalon and $\tau$ is the time for a wave to traverse the etalon once. Equation (6.21) allows us to write the phase delay across the etalon, instead of in distance $x$, in terms of frequency $f$, which is more convenient for studying frequency filters.

From electromagnetic theory, the electric field at the conducting mirror is zero, $E_{\text{mirror}} = 0$, so the field across the etalon $E(x) \propto \sin(\beta x)$, which requires $\beta x$ to be an arbitrary integer multiple of $\pi$. Hence, for $m$ an arbitrary integer,

$$\beta x = m\pi \quad \text{or} \quad x = \frac{m\pi}{\beta} \tag{6.22}$$

We conclude that resonance occurs only when the distance across the etalon $x$ is a multiple number of half-wavelengths $\lambda/2$ as shown in Figure 6.12 and by using $\beta = 2\pi/\lambda$ in equation (6.22),

$$x = m\lambda/2 \tag{6.23}$$

Transmission through the etalon is the output field, equation (6.19), divided by the input field, $E_i$, with $\beta x$ from equation (6.21):

$$H(f) = \frac{E_o(f)}{E_i(f)} = (1 - A - R)e^{-j2\pi f\tau} \sum_{m=0}^{\infty} \left[ R e^{-i4\pi f\tau} \right]^m \tag{6.24}$$

From tables or by using Maple, an infinite summation such as this can be written in closed form:

$$\sum_{m=0}^{\infty} a^m = \frac{1}{1-a} \quad \text{for} \quad a < 1 \tag{6.25}$$

Using equation (6.25) with equation (6.24) to condense the infinite series gives

$$H(f) = (1 - A - R)e^{-j2\pi f\tau} \frac{1}{1 - Re^{-j4\pi f\tau}} \tag{6.26}$$

The transfer function for power transmission through the cavity can be written as

$$
\begin{aligned}
T(\tau) = |H(f)|^2 &= H(f)H(f)^* \\
&= \frac{(1 - A - R)^2}{(1 - Re^{-j4\pi f\tau})(1 - Re^{j4\pi f\tau})} \\
&= \frac{(1 - A - R)^2}{1 + R^2 - 2R\cos 4\pi f\tau} \quad \text{from } e^{j\theta} + e^{-j\theta}/2 = \cos\theta \\
&= \frac{(1 - A - R)^2}{1 + R^2 - 2R(1 - 2\sin^2 2\pi f\tau)} \quad \text{from } \cos 2\theta = 1 - 2\sin^2\theta \\
&= \frac{(1 - A - R)^2}{(1 - R)^2 + R(2^2 \sin^2 2\pi f\tau)} \\
&= \frac{(1 - A/(1 - R))^2}{1 + ((2\sqrt{R}/(1 - R))\sin 2\pi f\tau)^2} \quad \text{dividing top and bottom by } (1 - R)^2
\end{aligned}
\tag{6.27}
$$

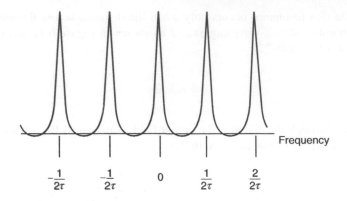

**FIGURE 6.14** Longitudinal frequency modes out of Fabry–Perot resonator.

### 6.2.2.2 Free Spectral Range

Equation (6.27) peaks for frequencies at which $\sin(2\pi f\tau) = 0$, computed from

$$2\pi f\tau = 0, \pi, 2\pi, 3\pi, \ldots \quad \text{or} \quad 2\pi f\tau = m\pi, \quad f = 0, \frac{1}{2\tau}, \frac{2}{2\tau}, \frac{3}{2\tau}, \ldots \quad (6.28)$$

where $\tau$ is one-way time to travel across etalon. Each peak corresponds to a laser longitudinal mode in Figure 6.14. Free spectral range (FSR) is the range of frequencies over which tuning is possible by moving one mirror away from the other until phase wrapping occurs to the next half-wavelength. Hence, from Figure 6.14,

$$\text{FSR} = \frac{1}{2\tau} = \frac{1}{2}\frac{c}{x}, \quad \text{for } n = 1 \quad (6.29)$$

Previously, we represented modes in terms of the number of half-cycles of wavelength that fit into the distance across the cavity width $x$ (equation (6.23) and Figure 6.12). In equation (6.28) and Figure 6.14, we represent modes as resonant frequencies separated by half the reciprocal of the travel time $\tau$ across the etalon.

*Maximum.* From equation (6.27), the maximum occurs at $\sin 2\pi f\tau = 0$ and is

$$T(f) = \left(1 - \frac{A}{1 - R}\right)^2 \quad (6.30)$$

Laser power transfer function $T(f)$ is very large because gain $A$ is high relative to $1 - R$, where $0 < R < 1$.

*Half-maximum power.* Half-maximum power occurs at frequency $f_1$ when denominator of equation (6.27) is two or long bracket is one.

$$\frac{2\sqrt{R}}{1 - R} \sin 2\pi f_1\tau = 1 \quad \text{or} \quad \sin 2\pi f_1\tau = \frac{1 - R}{2\sqrt{R}} \quad (6.31)$$

For a resonator, we generally have a small $2\pi f_1\tau$, so $\sin(2\pi f_1\tau) \to 2\pi f_1\tau$, or

$$f_1 = \left(\frac{1}{2\tau}\right)\frac{1-R}{\pi 2\sqrt{R}} \tag{6.32}$$

*Half-power bandwidth.* Half-power bandwidth (HPBW) is defined as

$$\text{HPBW} = 2f_1 = \left(\frac{1}{2\tau}\right)\frac{1-R}{\pi\sqrt{R}} \tag{6.33}$$

*Finesse.* Finesse is defined as

$$F = \frac{\text{FSR}}{\text{HPBW}} = \frac{1/2\tau}{((1/2\tau)/(1-R)/(\pi\sqrt{R}))} = \frac{\pi\sqrt{R}}{1-R} \tag{6.34}$$

Finesse indicates how many unique frequency channels may be placed inside the free spectral range before mode overlap arises. The number of available channels is one less than $F$ as deduced from examination of Figure 6.15, where although the finesse is three, the number of unique frequency channels is two because one of the three channels is split half at each end and therefore ambiguous in frequency. Hence, if we desire 100 channels, we need to design for 101. This is accomplished by increasing FSR by a factor 101/100. Furthermore, as the HPBW allows too much crosstalk between channels, it is normal to assume that we need 50–65% more channels and then use only the desired number of channels spaced out. Finesse also represents the average number of times a photon bounces back and forth in a cavity before the power in the cavity has fallen by $1/e$

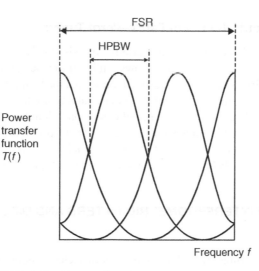

**FIGURE 6.15**　The number of unique channels from finesse.

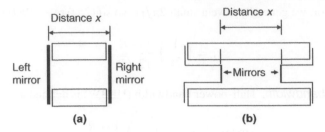

**FIGURE 6.16** Tuning a Fabry–Perot filter with piezoelectric material: (a) piezoelectric material inside etalon and (b) with mechanical gain.

*Contrast.* Contrast may be defined as max / min for the Fabry–Perot resonator output. From equation (6.27), the maximum and minimum values are

$$\text{Max} = \left(1 - \frac{A}{1-R}\right)^2$$

$$\text{Min} = \frac{(1-(A/(1-R)))^2}{1+(4R/(1-R)^2)}, \quad \text{where } \sin(2\pi f\tau) = 1$$

$$\text{Contrast} = \frac{\text{Max}}{\text{Min}} = 1 + \left(\frac{2\sqrt{R}}{1-R}\right)^2$$

Using equation (6.34),

$$\text{Contrast} = 1 + \left(\frac{2F}{\pi}\right)^2 \qquad (6.35)$$

### 6.2.3   Piezoelectric Tuning of Fabry–Perot Tuners

Piezoelectric material is convenient because it expands slowly with increasing voltage across its ends [53]. Figure 6.16a shows a tube of piezoelectric material placed between the two mirrors of a Fabry–Perot etalon so that the beam in the cavity passes through the hole in the tube. If piezoelectric material has length $X$ and the extension for a full range of voltage is $\Delta X$, the ratio of extension to length is $\Delta X/X = r$. Typically, $r = 0.05$. When this expansion is inadequate for providing a desired tuning range $R$, we use a mechanical gain of $G = R/r$ by making the piezoelectric material $G$ times longer than the Fabry–Perot cavity, as shown in Figure 6.16b.

## 6.3   THIN-FILM INTERFEROMETRIC FILTERS AND DIELECTRIC MIRRORS

Thin-film layered coatings are widely used for optical elements, especially with high-power lasers to minimize losses. Even small reflections from high-power laser beams can have enough power to damage eyes during testing. A stack of thin-film interferometric layers can filter a light beam.

### 6.3.1 Applications for Thin Films

#### *6.3.1.1 Dielectric Mirrors for a Single Frequency* High-power lasers can damage mirrors through $I^2 R_0$ loss in metal conducting surfaces, where for the conductor, $I$ is the induced current and $R_0$ is the resistance. Consequently, layers of dielectric material with no conducting materials, hence with no $I^2 R$ loss, are used in place of metal mirrors in high-power applications, such as the Airborne Laser program (Section 12.2). If the thickness of a layer is half the wavelength, $d = \lambda/2$, the reflection from the front side of the layer will be in phase with that from the rear side, which has traveled an extra distance of $2d = \lambda$. The amplitude of the reflectance will also depend on the refractive indices across the interfaces, equation (6.36).

As discussed in Section 6.2, for incident light of field $E_i$, if the power reflectivity is $R$, the reflected field will be $\sqrt{R}E_i$ and the transmitted field $\sqrt{1 - R}E_i$ because of conservation of power. If there are $N$ identical layers, the transmission through the stack of $N$ layers will be approximately $(\sqrt{1 - R})^N E_i$, which can be extremely small for large $N$. For infrared frequencies, the thickness of a layer is $d = \lambda/2 \approx 500$ nm so the thickness of 100 layers is still only 0.05 $\mu$m. The light that is not transmitted is reflected, showing that extremely high reflectivities are possible, much higher than that for a metal mirror. When almost the entire high-power beam is reflected, there is little heating effect from residual resistance and the mirror remains undamaged. A more exact computation can be made with the matrix method of Section 6.3.2.

Students in my laboratory designed and constructed an interferometer used for remote listening. The system involves shining a laser beam across a street onto a window of a conference room in another building. They used a small square panel of glass to represent the window. The reflections from the front and back sides of the window glass interfere. Conversation in the conference room vibrates the glass panel at audio rates. The bending of the glass changes the angle of incidence, so the thickness of the glass along the beam changes, causing the back reflected beam to change phase relative to the front reflected beam. The thickness has only to vary nanometers for adequate phase difference. The phase difference modulates the intensity of the light reflected back to a photodetector next to the laser. The detector cannot respond to the frequency of light and sees only the modulation. The output of the detector is fed to an audio amplifier from which conversation in the room across the street can be heard. Military and commercial secure rooms that have windows often have white noise generators in their windows to reduce the effectiveness of such a spy system.

Dielectric periodic structures in thin film or etched in relief into semiconductor materials, to act as a dielectric mirror, are used in distributed feedback lasers to reflect only a very narrow band of frequencies. This allows selection of a single longitudinal mode in a Fabry–Perot laser or resonator (Figure 6.13) for wavelength division multiplexing. The narrow laser frequency linewidth allows hundreds of channels of varying wavelength to be simultaneously amplified through a band-limited optical amplifier. Periodic structures are also etched into fiber for filtering in telecommunications, for example, for dispersion compensation or equalization every 40 km in optical fiber networks.

### 6.3.1.2 *Dielectric Mirrors and Filters for Multiple Frequencies* To protect eyes, military goggles and windshields highly reflect light at the major laser weapon frequencies (Section 8.1.3). In this case, a stack of thin films is used for each frequency. The resulting thin-film filters have such narrow bandwidth that the overall light intensity is not noticeably impacted (Chapter 14).

In another student project, a small low-cost portable system was developed for measuring the accumulated UV-A and UV-B that cause skin damage. A UV detector and integrator indicates when safe levels of sun exposure are exceeded. Phosphor thin films were used to convert the UV to visible light for low-cost silicon sensing [83, 99].

### 6.3.1.3 *Antireflection Coatings* In high-power laser systems, optical elements are coated with antireflection coating to minimize losses in the system. If a thin-film layer has thickness $d = \lambda/4$, light reflecting from the front of the layer is exactly $180°$ out of phase with light reflecting from the back that has traveled $2d = \lambda/2$ farther. Thus, the reflected wave cancels by destructive interference. Many layers can provide extremely low reflectance, in a manner similar to obtaining high reflectivity for dielectric mirrors. Layers of different widths can optically filter light coming in; for example, optical antireflection coatings on eyeglasses cover the visible spectrum with several layers. Layered media [18] arise in many fields, for example, by deposition in nature [80, 81], in integrated circuits in lithography [95, 97], and in thin-film filters for telecommunication coarse wavelength division multiplexing (CWDM).

Stealth vehicles are designed to give low radar or sonar signatures. This implies reducing diffraction (Chapter 3) which includes reflection. It also means removing sharp metal objects that diffract in all directions, making detection possible from any angle. Aerodynamic shapes, such as a sharp wing edge, can be achieved by molding composite materials around smooth metal shapes. Absorption is also used to suppress reflection. Recently, there has been increased interest in making objects invisible for optical or radar wavelengths. Metamaterials of periodic structures allow negative refractive index materials to bend light in the opposite direction to normal refractive index materials. Layered coatings on an object can now cause monochromatic electromagnetic waves to bend around an object and return to the path along which it originally came, so the object appears to have vanished. It is more difficult to make this work at optical frequencies for a broader band.

### 6.3.2 Forward Computation Through Thin-Film Layers with Matrix Method

We use the matrix method [45, 56, 156]. We have used it previously in geophysics [80, 81], for leaking fields in integrated optics [95], and for curved dielectric layered fields [97]. We have also used it for determining the values of refractive index in a

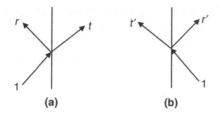

**FIGURE 6.17** Reflection and transmission of TE mode at an interface: (a) entering from left and (b) entering from right.

few layers when illuminating the waveguide at an angle and observing the angle of output beams for each mode (method of lines) as used in the prism coupling method in optical integrated circuits.

Figure 6.17 shows the reflection $r$ and transmission $t$ for a wave of unity amplitude entering from the left side of an interface (Figure 6.17a) and reflection $r'$ and transmission $t'$ for a unity amplitude wave entering from the right side of the interface (Figure 6.17b). We assume normal rays, so that reflection and transmission do not change the angles of the waves. Angles and absorption are included in Ref. [81].

Figure 6.18 shows propagation of right and left fields across a single layer, the $k$th layer. Superscript "+" refers to right moving fields, "−" to left side of an interface, and $k$ to the layer number. The reflection coefficients for the TE waveguide mode, having the electric field $E$ perpendicular to the plane of incidence, entering from the left and right sides of the $k$th interface are, respectively, given by

$$r_k = \frac{n_k - n_{k-1}}{n_k + n_{k-1}} \quad \text{and} \quad r'_k = -r_k \tag{6.36}$$

The transmission coefficients for the TE waveguide mode from the left and right sides of the $k$th interface are, respectively, given by

$$t_k = \frac{2n_k}{n_k + n_{k-1}} \quad \text{and} \quad t'_k = \frac{2n_{k-1}}{n_k + n_{k-1}} \tag{6.37}$$

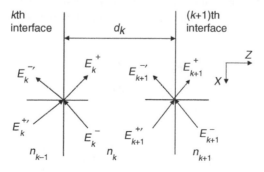

**FIGURE 6.18** Propagation across a single layer.

We derive an equation for the right and left fields entering the $k$th interface in Figure 6.18 in terms of those entering the $(k + 1)$th interface—in other words, the field $E_k'^+$, $E_k'^-$ just to the left of the $k$th interface to the field $E_{k+1}'^+$, $E_{k+1}'^-$ just to the left of the following $k + 1$th interface. Figures 6.17 and 6.18 show that

$$E_k^+ = t_k E_k'^+ + r_k' E_k^-$$
$$E_k'^- = t_k' E_k^- + r_k E_k'^+ \tag{6.38}$$

or

$$-t_k E_k'^+ \qquad = -E_k^+ + r_k' E_k^-$$
$$-r_k E_k'^+ + E_k'^- = \qquad t_k' E_k^-$$

Therefore, in matrix form,

$$\begin{bmatrix} -t_k & 0 \\ -r_k & 1 \end{bmatrix} \begin{bmatrix} E_k'^+ \\ E_k'^- \end{bmatrix} = \begin{bmatrix} -1 & r_k' \\ 0 & t_k' \end{bmatrix} \begin{bmatrix} E_k^+ \\ E_k^- \end{bmatrix} \tag{6.39}$$

By inverting the matrix on the left-hand side, this gives

$$\begin{bmatrix} E_k'^+ \\ E_k'^- \end{bmatrix} = \frac{1}{-t_k} \begin{bmatrix} 1 & 0 \\ r_k & -t_k \end{bmatrix} \begin{bmatrix} -1 & r_k' \\ 0 & t_k' \end{bmatrix} \begin{bmatrix} E_k^+ \\ E_k^- \end{bmatrix}$$

$$= \frac{1}{-t_k} \begin{bmatrix} -1 & -r_k \\ -r_k & r_k r_k' - t_k t_k' \end{bmatrix} \begin{bmatrix} E_k^+ \\ E_k^- \end{bmatrix} \tag{6.40}$$

where we used equation (6.36), $r_k r_k' - t_k t_k' = 1$ from power conservation. So,

$$\begin{bmatrix} E_k'^+ \\ E_k'^- \end{bmatrix} = \frac{1}{t_k} \begin{bmatrix} 1 & r_k \\ r_k & 1 \end{bmatrix} \begin{bmatrix} E_k^+ \\ E_k^- \end{bmatrix} \tag{6.41}$$

From Figure 6.18, in passing from $k$th layer to the $(k + 1)$th layer or from $(k + 1)$th layer to the $k$th layer, there is a unit delay represented by the $\sqrt{Z}$ transform operator (time to cross layer) [133]:

$$E_{k+1}'^+ = \sqrt{Z} E_k^+$$
$$E_k^- = \sqrt{Z} E_{k+1}'^- \tag{6.42}$$

or

$$\begin{bmatrix} E_k^+ \\ E_k^- \end{bmatrix} = \begin{bmatrix} 1/\sqrt{Z} & 0 \\ 0 & \sqrt{Z} \end{bmatrix} \begin{bmatrix} E_{k+1}'^+ \\ E_{k+1}'^- \end{bmatrix} \tag{6.43}$$

Substituting equation (6.43) into equation (6.41) gives the relation sought between just left of the $k$th interface to just left of the $(k + 1)$th interface.

$$
\begin{bmatrix} E_k'^+ \\ E_k'^- \end{bmatrix} = \frac{1}{t_k} \begin{bmatrix} 1 & r_k \\ r_k & 1 \end{bmatrix} \begin{bmatrix} 1/\sqrt{Z} & 0 \\ 0 & \sqrt{Z} \end{bmatrix} \begin{bmatrix} E_{k+1}'^+ \\ E_{k+1}'^- \end{bmatrix}
$$

$$
= \frac{1}{t_k} \begin{bmatrix} 1/\sqrt{Z} & r_k\sqrt{Z} \\ r_k/\sqrt{Z} & \sqrt{Z} \end{bmatrix} \begin{bmatrix} E_{k+1}'^+ \\ E_{k+1}'^- \end{bmatrix} \tag{6.44}
$$

From equation (6.44), we name the matrix for propagating through the $k$th layer from right to left as

$$
\mathbf{S}_k = \frac{1}{t_k} \begin{bmatrix} 1/\sqrt{Z} & r_k\sqrt{Z} \\ r_k/\sqrt{Z} & \sqrt{Z} \end{bmatrix} = \frac{1}{\sqrt{Z}t_k} \begin{bmatrix} 1 & r_k Z \\ r_k & Z \end{bmatrix} \tag{6.45}
$$

Figure 6.19 shows a horizontal stack of layers. Using equations (6.45) and (6.43), the field at left top is related to field at left bottom by

$$
\begin{bmatrix} E_1^{+\prime} \\ E_1^{-\prime} \end{bmatrix} = \Pi_{k=1}^K \mathbf{S}_k \begin{bmatrix} E_{K+1}^{+\prime} \\ E_{K+1}^{-\prime} \end{bmatrix} \tag{6.46}
$$

where $\Pi_{k=1}^K$ is the product of $2 \times 2$ matrices from $k = 1$ to $k = K$. There is a final interface, the $(K + 1)$th interface at the right, that can be accounted for by assuming an additional $(K + 1)$th layer with zero width $d_{K+1} = 0$ for which the scattering matrix, equation (6.45) or equation (6.41), reduces to

$$
\mathbf{S}_{K+1} = \frac{1}{t_{K+1}} \begin{bmatrix} 1 & r_{K+1} \\ r_{K+1} & 1 \end{bmatrix} \tag{6.47}
$$

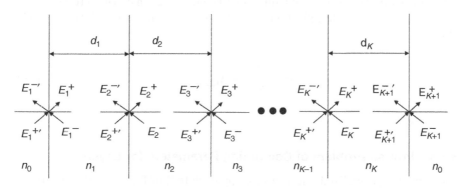

**FIGURE 6.19** Horizontal layer stack.

Then

$$
\begin{bmatrix} E_{K+1}^{+\prime} \\ E_{K+1}^{-\prime} \end{bmatrix} = \frac{1}{t_{K+1}} \begin{bmatrix} 1 & r_{K+1} \\ r_{K+1} & 1 \end{bmatrix} \begin{bmatrix} E_{K+1}^{+} \\ E_{K+1}^{-} \end{bmatrix} = S_{K+1} \begin{bmatrix} E_{K+1}^{+} \\ E_{K+1}^{-} \end{bmatrix} \tag{6.48}
$$

We can define the product result for $2 \times 2$ matrices for the whole plane layer stack in a matrix $\mathbf{M}$ with elements $M_{11}$, $M_{12}$, $M_{21}$, and $M_{22}$:

$$
\begin{bmatrix} E_1^{+\prime} \\ E_1^{-\prime} \end{bmatrix} = S_{K+1} \Pi_{k=1}^{K} S_k \begin{bmatrix} E_{K+1}^{+} \\ E_{K+1}^{-} \end{bmatrix} = \begin{bmatrix} M_{11} & M_{12} \\ M_{21} & M_{22} \end{bmatrix} \begin{bmatrix} E_{K+1}^{+} \\ E_{K+1}^{-} \end{bmatrix} \tag{6.49}
$$

In equation (6.49), there are four unknowns, the right and left fields at the top of the stack of layers and the right and left fields at the output at the right of the stack, but only two equations. The specific problem determines which of these four unknowns can be assumed to be known or zero. For example, if we assume that there are no further reflections after the $(K + 1)$th one at the right, then no field couples back into the right side of the layer stack and $E_{K+1}^{-} = 0$. Then equation (6.49) can be written as

$$
\begin{bmatrix} E_1^{+\prime} \\ E_1^{-\prime} \end{bmatrix} = \begin{bmatrix} M_{11} & M_{12} \\ M_{21} & M_{22} \end{bmatrix} \begin{bmatrix} E_{K+1}^{+} \\ 0 \end{bmatrix} \tag{6.50}
$$

We can now compute transmittance $Q$ for the stack $E_{K+1}^{+}$ relative to the input $E_1^{+\prime}$ by dividing one by the other:

$$
Q = \frac{E_{K+1}^{+}}{E_1^{+\prime}} = M_{11}^{-1} \tag{6.51}
$$

where the series $M_{11}$ is a single term of matrix $\mathbf{M}$, defined in equation (6.49).

We can also compute reflection from the stack $E_1^{-\prime}$ in terms of the input $E_1^{+\prime}$ [80, 81]:

$$
P = \frac{E_1^{-\prime}}{E_1^{+\prime}} = M_{22} M_{11}^{-1} \tag{6.52}
$$

### 6.3.3 Inverse Problem of Computing Parameters for Layers

Following the generalized linear inverse approach [80] of Figure 6.20, we compute a Jacobian (or sensitivity) matrix that shows the sensitivity of the output, that is, the

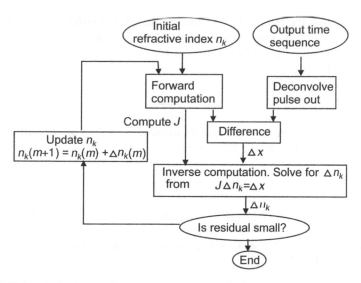

**FIGURE 6.20**  Block diagram for inverse computation for inspection of unopened packages.

effect on output $Q_p$ at time sample $p$ due to a change $n_k$ in the $k$th layer refractive index.

$$J_{p,k} = \frac{\partial Q_p}{\partial n_k} \tag{6.53}$$

From equation (6.51), $Q_p$ results from the product of $2 \times 2$ matrices for each layer. From equation (6.36), for the $k$th layer, only $n_k$ and $n_{k-1}$ are involved, the other matrices in the $2 \times 2$ product chain remaining unchanged [80]. In this case, an element of the Jacobian matrix can be computed from

$$
\begin{aligned}
J_{p,j} &= \left[\Pi_{k=1}^{j}{}^{2}S_k\right] \left\{ \frac{\partial(M_{j-1}M_j)}{\partial n_j} \right\} \left[\Pi_{k=j+1}^{K+1}S_k\right] \\
&= \left[\Pi_{k=1}^{j-2}S_k\right] \left\{ M_{j-1}\left(\frac{\partial M_j}{\partial n_j}\right) + \left(\frac{\partial M_{j-1}}{\partial n_j}\right)M_j \right\} \left[\Pi_{k=j+1}^{K+1}S_k\right] \\
&= \left[\Pi_{k=1}^{j-1}S_k\right] \left\{ \frac{\partial M_j}{\partial n_j} \right\} \left[\Pi_{k=j+1}^{K+1}S_k\right] \\
&\quad + \left[\Pi_{k=1}^{j-2}S_k\right] \left\{ \frac{\partial M_{j-1}}{\partial n_j} \right\} \left[\Pi_{k=j}^{K+1}S_k\right]
\end{aligned} \tag{6.54}
$$

From equation (6.54), the four product terms are partial computations for the forward computation. So the Jacobian is most easily computed at the same time as the forward computation (Figure 6.20).

Using a Gauss–Newton approach, the output of the forward computation series $Q_p$ as a function of time is now compared with that measured at the sensor $Q'$

after deconvolving the pulse out if necessary. Hence, the difference vector $\Delta y_p = Q_p - Q'_p$ is computed. The difference vector $\Delta y$ is now used with the Jacobian matrix $J_{p,j}$ to compute an update vector $\Delta n$ for the refractive indices for the layers by solving the linear equation

$$\Delta \mathbf{y} = \mathbf{J} \Delta \mathbf{n} \tag{6.55}$$

The refractive indices are now updated for the next $(m + 1)$th iteration of the generalized inverse algorithm (Figure 6.20) using

$$n_{m+1} = n_m + \Delta n_k(m) \tag{6.56}$$

We repeat loops of forward and inverse computations until the RMS of the update vector converges to a lower limit. When the Jacobian is sparse, we use a conjugate gradient algorithm [83]. The method was successfully applied and adopted in studies in geophysics [80, 81].

# LASER TECHNOLOGY FOR DEFENSE SYSTEMS

# CHAPTER 7

# PRINCIPLES FOR BOUND ELECTRON STATE LASERS

The word *laser* is derived from the acronym *light amplification by stimulated emission of radiation* and the name *maser* from the acronym *microwave amplification by stimulated emission of radiation*. Stimulated emission refers to Bremsstrahlung radiation, converting energy produced from showing down of electrons into microwave or light radiation.

For coherent electromagnetic radiation, below 1 GHz we have radio frequencies that are lower than microwaves and for which masers are not generally used. From 1 to 100 GHz, we have microwaves that can be generated with masers (some use 300 GHz as the upper limit). Between 100 and 300 GHz, radiation is sometimes called millimeter waves and falls between masers and lasers. Above 300 GHz, we have light that can be generated with lasers. Higher frequencies with more energy per photon involve X-rays, gamma rays, nuclear rays, and cosmic rays. X-ray lasers have also been demonstrated and applied to high-resolution VLSI lithography.

Bremsstrahlung means slowing down of electrons in an electron beam to convert energy to electromagnetic radiation. There are two types of bremsstrahlung stimulated emission. This chapter discusses the first type, that is, lasers that generate coherent photons from electrons jumping from a higher energy to a lower energy bound electron state. Frequency is determined by the material energy levels. The second type of bremsstrahlung stimulated emission is the cyclotron maser or laser (free-electron

*Military Laser Technology for Defense: Technology for Revolutionizing 21st Century Warfare*,
First Edition. By Alastair D. McAulay.
© 2011 John Wiley & Sons, Inc. Published 2011 by John Wiley & Sons, Inc.

laser), described in Chapter 10 and 11, which is used for ultrahigh-power masers and lasers at any frequency.

In Section 7.1, we list some of the significant advantages of coherent light that help revolutionize warfare and the basic light–matter interaction process for generating laser light in bound electron state lasers. In Section 7.2, we describe the basic semi-conductor laser diode widely used for low-power applications, such as pumps for more powerful lasers and in arrays. In section 7.3, we discuss semiconductor optical amplifiers.

## 7.1   LASER GENERATION OF BOUND ELECTRON STATE COHERENT RADIATION

We discuss the advantages of coherent radiation and then the basic light–matter interaction for bound electron state lasers.

### 7.1.1   Advantages of Coherent Light from a Laser

Coherent light has temporal coherence (Section 6.1), and spatial coherence (Sections 1.3.6 and 3.2.2). Some advantages of coherent light radiation are discussed in this section from the system point of view.

1. Coherent light has the ability to project energy over a distance, such as for a laser pointer, a laser designator, a range finder, or a laser weapon (Chapter 12).

2. Coherent light can be focused to a small region for a weapon or for dense compact optical disk storage.

3. Coherent light can detect chemical and biological weapons because the wavelength is comparable to the size of molecules and bacteria.

4. Frequencies of around $10^{14}$ Hz provide the potential to send $10^{14}$ bit/s, one cycle is on or off, which corresponds to a hundred thousand billion bits per second through air or in fiber. This is important for optical communications and the optical fiber-connected Internet.

5. The short wavelength allows high-resolution imaging for satellites and operation with day or night vision cameras.

A significant advantage of coherent radiation is that the energy is close to a single frequency that produces an entropy (or uncertainty) approaching zero: there is no uncertainty in a sine wave. In contrast, thermal energy from burning coal, or to some extent from the sun, is disordered with high entropy. When coherent light enters a system, it lowers the system entropy, whereas when a system is heated, it increases the entropy. Higher entropy lowers efficiency. In a chemical laser, the heat generated in a chemical reaction is used to pump a laser directly to create coherent electromagnetic waves. Direct generation of high-power coherent light can be efficient in generation (Chapter 10) and distribution, especially in future, as some predict that energy will be used mainly for lighting and computing [83]. This suggests that new approaches for energy generation and distribution may be possible.

### 7.1.2 Basic Light–Matter Interaction Theory for Generating Coherent Light

Creating coherent light radiation for bound electron states, at close to a single frequency in the region visible light (300 nm to 5μm), can be described by photons, based on Einstein's 1917 light–matter interaction research, Planck's spectral distribution of blackbody radiation law, and Boltzmann's statistics on the distribution of atomic population between different energy levels in an atom [176].

We show that a pump and a resonator are required to create coherent light radiation in a bound electron state laser. In such a state, electrons around an atom exist in only one of a number of states of different energy levels according to quantum mechanics. Although three or four states are required for lasing, we first consider transitions between energy levels for a simpler two-state case for which there is an energy density of $N_1$ atoms per unit volume in the lower state $E_1$ and an energy density of $N_2$ atoms per unit volume in the upper state $E_2$ [67, 119]. Good lasing materials have sharp fluorescent lines, strong absorption bands, and high quantum efficiency for fluorescent transitions of interest. Furthermore, efficient lasers use direct bandgap materials such as gallium arsenide (GaAs) for which the dip in energy in the upper energy level occurs at the same electron momentum as the peak in the lower energy level. In such a material, when an electron (or carrier) in an atom (and consequently the energy in the atom) falls from an energy level $E_2$ to a lower energy level $E_1$ by crossing the bandgap $E_g$, a photon of light is emitted of energy corresponding to the energy loss of an electron:

$$E_g = E_2 - E_1 = hv = \hbar\omega \tag{7.1}$$

where $h$ is Plank's constant, $v$ is the frequency, $\hbar = h/2\pi$, and $v = 2\pi\omega$. Different materials have different bandgaps; consequently, they will emit different frequencies (or colors) of light.

Einstein's theory of light–matter interaction explains the following ways for electrons to move between energy levels [67, 119] (Figure 7.1).

1. *Absorption*: If light with a photon of higher energy than $v$ in equation (7.1) falls on the material, the photon is absorbed by an electron jumping across the bandgap $E_g$, from the lower $E_1$ to the higher $E_2$ energy level. The energy from the photon is stored at the higher energy level. For optical pumping, as described here, the number of atoms transitioning from lower to higher energy level depends on the flux density of input light radiation $D(v)$ and the atomic density $N_1$, according to

$$R_{abs} = \frac{dN_1}{dt} = -B_{12}D(v)N_1 \tag{7.2}$$

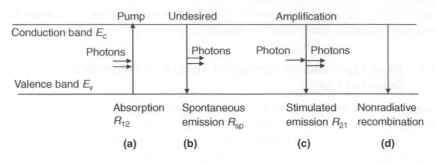

**FIGURE 7.1**   Energy-level transitions.

where $B_{12}$ is the Einstein absorption coefficient (or proportionality absorption constant), and $v$ is the frequency of the input light. When $N_2$ is greater than $N_1$, we have population inversion, which is required for lasing. Absorption also enables photodetection, where an absorbed photon generates an electric current from low to high energy levels.

2. *Spontaneous Emission of Photons*: Electrons emit photons when falling randomly from a higher energy level to a lower one without external influence. These photons are incoherent with each other, having random polarization, phase, and frequency. Over time, spontaneous emission lowers the upper level carrier density $N_2$, the number of carriers per unit volume, to below that for the lower energy level $N_1$ according to Boltzmann's statistics

$$\frac{N_2}{N_1} = \exp\left\{\frac{E_g}{k_B T}\right\} = \exp\left\{\frac{-hv}{k_B T}\right\} \tag{7.3}$$

where $k_B$ is Boltzmann's constant and $T$ is the absolute temperature. At room temperature (293K), $N_2 < N_1$ and lasing cannot occur because of lack of carriers in the upper level. Consequently, pumping with light or electric current is needed to enable lasing. The rate of transition of carriers from upper $E_2$ to lower $E_1$ level by spontaneous transition is

$$R_{spon} = \frac{dN_2}{dt} = A_{21} N_2 \tag{7.4}$$

where $A_{21}$ is the Einstein coefficient (or proportionality constant) representing probability of spontaneous transitions from $E_2$ to $E_1$.

3. *Stimulated Emission of Photons*: An incident photon stimulates a laser medium that has gain and a population inversion $N_2 > N_1$, causing an electron to fall from energy level $E_2$ to $E_1$. In the process, a photon is emitted of energy $E_g = E_2 - E_1$ that is indistinguishable from the stimulating photon: same directional properties, same polarization, same phase, and same spectral characteristics. The properties of coherent light result from a large number of photons

in coherent lockstep. The rate of stimulated emissions is

$$R_{\text{stim}} = \frac{dN_2}{dt} = B_{21} N_2 D(v) \tag{7.5}$$

where $B_{21}$ is the Einstein coefficient for stimulated emission and $D(v)$ is the flux density of light in the cavity at frequency $v$.

4. *Nonradiative Deexcitation*: An electron can fall from a higher to a lower energy level without generating a photon. The energy lost by the carrier appears in other forms, such as translational, vibrational, or rotational.

Assuming that $N_1$ and $N_2$ are the electron densities in lower and upper energy states, from equations (7.2), (7.4) and (7.5), the up and down rates between the two states must be equal at thermal equilibrium:

$$\overbrace{A_{21} N_2}^{\text{Spontaneous emission}} + \overbrace{B_{21} N_2 D(v)}^{\text{Stimulated emission}} = \overbrace{B_{12} D(v) N_1}^{\text{Absorption}} \tag{7.6}$$

Using Boltzmann's equation, equation (7.3), to replace $N_2/N_1$ in equation (7.6) and with the blackbody radiation law from [176]

$$D(v) = \frac{8 \pi n^3 h v^3}{c^3 (\exp\{h v/(k_B T) - 1\})} \tag{7.7}$$

gives Einstein's relation between $A$'s and $B$'s [176]:

$$\overbrace{A_{21}}^{\text{Spontaneous coefficient}} = \frac{8 \pi n^3 h v^3}{c^3} \overbrace{B_{21}}^{\text{Stimulated coefficient}} \tag{7.8}$$

and

$$\overbrace{B_{21}}^{\text{Absorption coefficient}} = \overbrace{B_{12}}^{\text{Stimulated coefficient}} \tag{7.9}$$

where the speed of light in material of refractive index $n$ is $c = c_0/n$, with $c_0$ the speed of light in vacuum or air. Equation (7.8) shows that the Einstein coefficient for the rate of spontaneous emission, causing the incoherent noise, is proportional to the Einstein coefficient for the rate of stimulated emission, the desired coherent light. So we cannot eliminate noise from the laser with useful output. Equation (7.9) shows that the Einstein coefficient for the rate of stimulated emission and that for the rate of absorption (or pumping) are equal. Consequently, two properties are required for lasing in a suitable material.

The first criterion for lasing is that the rate of stimulated emission generating laser light must be greater than the rate of absorption (otherwise all the light generated is

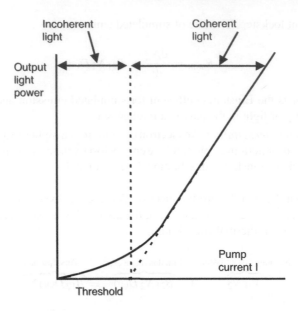

**FIGURE 7.2**    Laser characteristics illustrating threshold.

absorbed immediately). From equations (7.2) and (7.5),

$$\frac{|R_{stim}|}{|R_{abs}|} = \frac{B_{21}N_2 D(v)}{B_{12}N_1 D(v)} = \frac{N_2}{N_1} > 1 \tag{7.10}$$

When $N_2 > N_1$, there is inversion of electrons and lasing can occur by stimulated emission. Inversion does not naturally occur at room temperature, equation (7.3). So a laser has to be pumped with light having photons of higher energy than the material level bandgap. Flash tubes, fluorescent, and other incoherent lights can be used for pumping. The pump light must have sufficient power to overcome absorption. Below this threshold, the laser emits incoherent light and above the threshold lasing occurs and the light power out, $P_{out}$, is coherent and much greater (Figure 7.2). In the case of semiconductor lasers, pumping is performed with electric current $I$ rather than light. For pumping to be effective, the time constant associated with decay from the upper to lower levels must be longer than the recombination time so that a reserve of excess electrons is built up in the higher energy band.

The second criterion for generating laser light is that the stimulated emission rate must be greater than the spontaneous emission rate from equations (7.5) and (7.4) (otherwise incoherent noise light can overwhelm coherent laser light).

$$\frac{R_{stim}}{R_{spon}} = \frac{B_{21}N_2 D(v)}{A_{21}N_2} = \frac{B_{21}}{A_{21}} D(v) \tag{7.11}$$

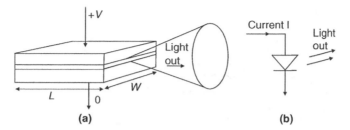

**FIGURE 7.3**   Laser diode: (a) structure and (b) symbol.

For equation (7.11), we need a high concentration of photons $D(v)$ in the laser. This is achieved with a resonant cavity that may be formed by means of two parallel mirrors or reflectors (Section 6.2), or an integrated optic ring resonator [96, 102, 105, 106]. In summary, we need a pump to provide carrier inversion, equation (7.10), and a resonant cavity to make sure that coherent light surpasses incoherent light, equation (7.11).

## 7.2   SEMICONDUCTOR LASER DIODES

Semiconductor laser diodes convert electric current into a beam of coherent light. They are efficient (over 50%), small (0.5 mm), and robust and are used as sources in long-haul telecommunication, compact disk players, laser pointers, and integrated and fiber-optic sensors, as well as for pumping solid-state lasers in military designator and range finding systems. Groups of laser diodes may be used for laser weapon systems (Section 7.2.4). Figure 7.3a shows a simple structure. A thin undoped intrinsic layer of a direct bandgap material such as GaAs is sandwiched between n-doped (excess electrons) and p-doped (reduced electrons) layers to form a p–n junction that acts as a diode. The sandwiched active layer is made from a slice of an approximately 30 cm diameter GaAs crystal which has been diced into thousands of small millimeter-size rectangular pieces, one for each laser diode. The cleaved faces at left and right sides of a piece reflect sufficient light to act as mirrors that form a cavity to bounce photons back and forth inside the laser diode. An electric current flows down through the junction to pump electrons in the intrinsic layer to a higher energy level for inversion. Light is emitted at the right as shown in Figure 7.3a. Figure 7.3b shows the symbol for a laser diode. Coherent light is emitted when forward current exceeds the value needed to overcome absorption.

### 7.2.1   p–n Junction

Figure 7.4 shows the energy level versus the momentum $k$ (wave vector for electron). The upper energy level (conductance level c), the n-doped level, holds excess electrons. The Fermi level $E_{fc}$ is the level below which the probability of finding an electron is greater than a half. Increasing the doping level will raise $E_{fc}$. The lower energy level (valence level v), the p-doped level, holds excess holes (absence of

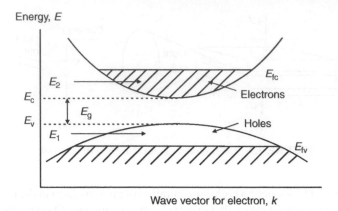

**FIGURE 7.4**    Semiconductor laser diode energy band diagram.

electrons). The Fermi level $E_{fv}$ is the level above which the probability of finding a hole is greater than 0.5. The least difference in energy levels over the wave number $k$ corresponds to the bandgap energy $E_g$ that determines the frequency of the light from equation (7.1). For pumping, the frequency (and hence energy) of the photons must be sufficient to exceed the bandgap energy $E_g = E_c - E_v$.

When the p-doped material and the n-doped material are brought together (Figure 7.5) for a p–n junction, the Fermi levels line up as shown. The result is an energy barrier $eV_0$ electron volts, where $e$ is the charge on an electron, preventing electrons from flowing from the n-doped material to the p-doped material.

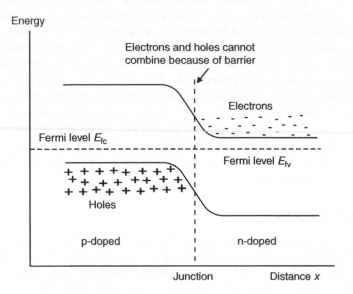

**FIGURE 7.5**    Semiconductor laser diode with no bias.

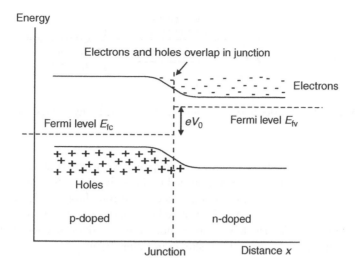

**FIGURE 7.6**  Forward-biased semiconductor laser diode.

By applying a forward voltage bias $V_f$ across the diode from a battery or a source (Figure 7.6), the barrier is reduced from $e(V_0)$ to $e(V_0 - V_f)$ and electrons in the n-doped material now orient directly above the holes in the so-called depletion width. Electrons can now readily transfer from higher to lower energy levels.

Figure 7.7 shows how activity is encouraged to focus at the junction in a double heterostructure laser diode. Figure 7.7a shows the schematic symbol for a laser diode. To confine carriers (electrons and holes) in the vicinity of the junction we sandwich a thin layer, called the active or intrinsic "i" layer, between the n- and p-doped materials to form a heterojunction or double heterostructure (Figure 7.7b). This improves performance by confining transfers between layers and by trapping light in the active

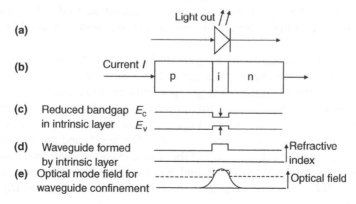

**FIGURE 7.7**  Double heterostructure laser diode: (a) symbol, (b) structure, (c) confinement with reduced bandgap, (d) refractive index for waveguide confinement, and (e) optical mode field for waveguide confinement.

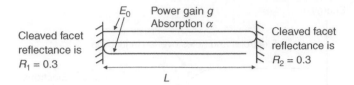

**FIGURE 7.8**    Steady-state condition in laser diode cavity at threshold.

layer. First, the active layer has a smaller bandgap than elsewhere (Figure 7.7c), which allows transfer between layers to be confined to the active layer. Second, by making the refractive index in the active layer larger than the surrounding n- and p-doped layer refractive indices, the light generated is trapped in an index guiding waveguide [130], as shown in Figure 7.7d. The lowest order mode field for the generated light shows that light is trapped in the active layer shown in Figure 7.7e.

### 7.2.2    Semiconductor Laser Diode Gain

The cleaved facets of the semiconductor laser diode act as reflectors for a Fabry–Perot resonator (Section 6.2), having a power reflectivity of

$$R_{\mathrm{m}} = \left( \frac{n-1}{n+1} \right)^2 \tag{7.12}$$

where we assumed refractive index $n = 1$ for air. For $n = 3.5$, a typical value of refractive index for GaAs, power reflectivity is $R_{\mathrm{m}} = 30\%$. The gain of the lasing material is sufficient to overcome the cavity loss at the imperfect mirrors. The reflectivity of the rear facet is often enhanced with optical reflective coating leaving only sufficient light for a photodetector to provide amplitude feedback control. The pump must provide sufficient current $I$ to overcome losses at the threshold for lasing, (Figures 7.2).

We now develop the equations for gain and resonance by taking an approach different from that used in Section 6.2, achieving results in a more illustrative form for our purpose [176]. If $g$ is the power gain per unit length in the active material and $\alpha$ is the internal absorption per unit length, we can write for a round-trip between facets in the cavity of length $L$ [176] (Figure 7.8) the amplitude gain (factor 2 for two-way and square root for amplitude) as $[e^{g2L}]^{1/2} = e^{gL}$ and the absorption as $[e^{-\alpha 2L}]^{1/2} = e^{-\alpha L}$. The light bounces off the two facets with power reflectivities $R_1$ and $R_2$ so that the amplitude loss for a round-trip is $\sqrt{R_1 R_2}$. The phase change during a round-trip is $2kL$. Therefore, for a steady state, the field $E_0$ after a round-trip equals that at the start:

$$E_0 = E_0 \sqrt{R_1 R_2} e^{gL} e^{-\alpha L} e^{j2kL} \tag{7.13}$$

Equating amplitude, leaving out the phase term, on each side of equation (7.13) gives

$$1 = \sqrt{R_1 R_2} e^{gL} e^{-\alpha L} \tag{7.14}$$

Taking natural logarithms

$$gL = \ln \left[ (R_1 R_2)^{-1/2} \right] + \alpha L \tag{7.15}$$

Only the active layer has gain material. So if we assume a confinement factor $\gamma = d/D < 1$, where $d$ is the thickness of the active layer and $D$ is the width of the lowest order mode (normally used), the effective gain per unit length is $g(d/D)$, less because some of the field is not in the active layer. So we can write gain equals loss as

$$g\frac{d}{D} = \alpha + \frac{1}{L}\ln \left[ (R_1 R_2)^{-1/2} \right] \quad \text{or} \quad g\frac{d}{D} = \alpha + \frac{1}{2L}\ln\frac{1}{R_1 R_2} \tag{7.16}$$

where loss in the cavity has attenuation $\alpha$ due to material absorption and second term $L^{-1}\ln[(R_1 R_2)^{-1/2}]$ due to laser light emitted from the cavity through the mirrors as output.

Equating phase in equation (7.13) gives $0 + jm2\pi = 2jkL$ or $kL = m\pi$. Using propagation constant $k = (2\pi)/(\lambda/n)$, the result in air, $n = 1$, is identical to that in equation (6.23):

$$L = m\frac{\lambda/n}{2} \tag{7.17}$$

which shows that the longitudinal mode $m$ is the number of half-wavelengths in the medium that fit across the cavity (Section 6.2, figure 6.12).

Using equation (6.29), the spacing of modes in wavelength $x_{\text{sp}}$ and in frequency $\nu_{\text{sp}}$ are then, respectively,

$$x_{\text{sp}} = (\lambda/n)/2 \quad \text{and} \quad \nu_{\text{sp}} = \frac{c/n}{2L} = \frac{1}{2\tau} \tag{7.18}$$

where $\tau = L/(c/n)$ is the travel time across the etalon. Hence, frequency spacing between nodes is again shown be equal to the reciprocal of the two-way travel time of light between the facets equation (6.28) (Figure 6.14).

As only modes with sufficient gain, above $g'$, are strong enough to overcome losses and will resonate, we exclude the small modes at either end in Figure 7.9 compared to Figure 6.13.

The gain of the active material in the laser diode at the nominal wavelength $\lambda_0$ depends on the injected pump current density, $J$ A/m$^2$, and other parameters

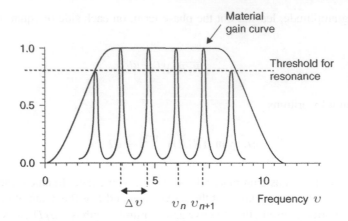

**FIGURE 7.9**  Longitudinal modes of frequency in a laser diode.

according to [57, 176]

$$g = \frac{\eta_q \lambda_0^2 J}{8\pi e n^2 d \Delta \nu} \tag{7.19}$$

where $\eta_q$ is the internal quantum efficiency for converting electrons to photons, $e$ is charge of an electron, $n$ is the refractive index of active layer, $d$ is the active layer thickness and $\Delta \nu$ is the linewidth of the laser output. The current needed to initiate lasing in the steady state, the threshold current density, $J_{th}$, occurs when gain equals loss. So substituting $g$ from equation (7.19) into equation (7.16) and solving for $J_{th}$ gives ($J > J_{th}$ for lasing)

$$J_{th} = \frac{8\pi e n^2 \Delta \nu D}{\eta_q \lambda_0^2} \left( \alpha + \frac{1}{L} \ln \frac{1}{R_1 R_2} \right) \tag{7.20}$$

Across the active layer in the diode, absorption loss can be written as $P_{loss} = \alpha P z$ and gain can be written as $P_{gain} = g P z$. Therefore, an internal efficiency can be written (with $g$ modified for confinement) as

$$\eta = \frac{P_{gain} - P_{loss}}{P_{gain}} = \frac{g(d/D) - \alpha}{g(d/D)} \tag{7.21}$$

or using equation (7.16) as

$$\eta = \frac{(1/2L)\ln(1/R_1 R_2)}{\alpha + (1/2L)\ln(1/R_1 R_2)} \tag{7.22}$$

The optical power into the laser diode is

$$P_{opt-in} = \frac{J}{e}(L \times W)\eta_q h\nu \tag{7.23}$$

where $J$ is the current density, $J/e$ is the number of electrons per second per area, $L \times W$ is area of active layer of length $L$ and width $W$, $\eta_q$ is the number of photons produced per electron, and $h\nu$ is the energy per photon. The optical power out of the laser diode is, using efficiency $\eta$ from equation (7.22),

$$P_{opt\ out} = \eta P_{opt-in} - \frac{(1/2L)\ln(1/R_1 R_2)}{\alpha + (1/2L)\ln(1/R_1 R_2)} \left(\frac{J}{e}\eta_q(L \times W)h\nu\right) \tag{7.24}$$

The electrical input power $P_{elec-in}$ differs from the optical input power $P_{opt-in}$ by including $I^2 R$ loss to get current onto the chip, in place of $\eta_q$ in equation (7.23):

$$P_{elec-in} = \frac{J}{e}(L \times W)h\nu + [J(L \times W)]^2 R_{series} \tag{7.25}$$

So the total so-called wall plug efficiency is obtained from equations (7.24) and (7.25) by

$$\eta_{wall} = \frac{P_{opt-out}}{P_{elec-in}} \tag{7.26}$$

## 7.2.3  Semiconductor Laser Dynamics

The simplest dynamic model of a bound electron state laser involves two first-order differential equations, called rate equations, that describe the interaction between two reservoirs: one for the electronic carrier density, $N$ carriers or electrons per cubic meter, and the other for the photon density, $N_{ph}$ photons per cubic meter. The interaction is complicated because the two reservoirs have different time constants.

For a laser diode, the level in the carrier density reservoir increases in time with the number of carriers per cubic meter $J/qd$ injected by the pump current density $J$ A/m$^2$, where $q$ is the charge on an electron and $d$ is the thickness of the active layer.

The level in the carrier density reservoir lowers in time due to recombination for stimulated emission that creates photons (Section 7.1.2). The stimulated emission rate $\Gamma v_g g(N) N_{ph}$ depends on the confinement factor $\Gamma$, (Section 7.2.2) group velocity $v_g$, the gain $g(N) = a(N - N_0)$, which is proportional to the excess carriers $N - N_0$ in the reservoir, where $N_0$ is that needed at threshold (transparency), and the number of photons in the photon reservoir $N_{ph}$. There is also a loss due to recombination that does not contribute to the generation of photons, $N/\tau_e$, where $1/\tau_e$ is the effective recombination rate. This can be written [75] as

$$\frac{\partial N}{\partial t} = \frac{J(t)}{qd} - \Gamma v_g g(N) N_{ph} - \frac{N}{\tau_e(N)} \tag{7.27}$$

The level in the photon density reservoir increases in time by the amount of the modified stimulated emission reduction $\Gamma v_g g(N) N_{ph}$ from the carrier density reservoir and decreases due to a photon decay time constant $\tau_{ph}$ that represents the time a photon stays in the cavity (longer for high-reflectivity mirrors). It also increases because a percentage $\beta_{sp}$ of the modified spontaneous emission rate $BN^2$, equation (7.4), is coherent with the stimulated emission. This can be written as

$$\frac{\partial N_{ph}}{\partial t} = \Gamma v_g g(N) N_{ph} - \frac{N_{ph}}{\tau_{ph}} + \beta_{sp} BN^2 \tag{7.28}$$

In Refs [98, 109], we show that the laser diode rate equations for a single optically injected laser diode, modeled by two coupled first-order equations as above, give rise to supercritical Hopf bifurcations that exactly mimic the popular Wilson–Cowan mathematical model for representing a neuron in the brain. The Wilson–Cowan model may be viewed as two cross-coupled dynamical nonlinear neural networks, one excitatory and the other inhibitory. Varying an input parameter, the sum of input intensities from all other incoming neurons, causes the Wilson-Cowan neural oscillator to move through a supercritical Hopf bifurcation so as to switch its output from a stable off state when the input is below a firing threshold to a stable oscillation state (limit cycle) for signals above the threshold, the frequency of which depends on the level of input stimulation. We note that a single optically injected laser diode was shown in Ref. [171] to be able to represent any form of nonlinear dynamics and chaos [153].

### 7.2.4 Semiconductor Arrays for High Power

Because of robustness, small size, and high efficiency, laser diodes are pervasive in the military for low-power applications. Higher power can be achieved by combining several broad-area laser diode emitters in a row on one substrate, called a bar. For still higher power, twenty 50 W laser bars of size 10 mm × 1 mm are arranged to form a 1 kW laser shown in Figure 7.10 [6]. Combining requires skill to cost effectively merge the beams to produce a single spatially coherent beam, and methods of maintaining beam quality and performance characteristics are discussed in Ref. [6].

## 7.3   SEMICONDUCTOR OPTICAL AMPLIFIERS

For any laser that generates coherent light, a variation can be made that amplifies light of the same wavelength. We illustrate this with a semiconductor optical amplifier (SOA) that is a variation of a laser diode. Normally, for high-power lasers (Section 8.2 and Chapter 13), optical amplifiers are used to boost power in several stages. For

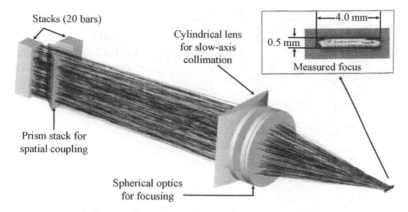

**FIGURE 7.10** Ray trace for 1 kW laser diode array.

light detection and ranging (lidar), (Chapter 15), and radio frequency detection and ranging (radar), including millimeter radar, (Chapter 16), complex signals are needed for optimal signal processing performance, so such signals are created electronically. They are then converted to optics when appropriate, for example, with a laser diode, and then amplified to high power with optical amplifiers.

A semiconductor optical amplifier (SOA) differs from a laser diode in that it avoids resonance. The reflecting facets of the semiconductor laser chip are coated with antireflection coatings (Section 6.3.1.3). If remnants of the resonance still show up noticeably, the SOA is known as a Fabry–Perot SOA. If the resonances are virtually eliminated, the SOA is known as a traveling-wave SOA. To further reduce resonance between front and rear facets, the waveguide is angled as shown in Figure 7.11.

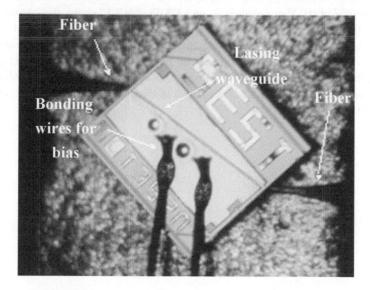

**FIGURE 7.11** Photograph of semiconductor optical amplifier.

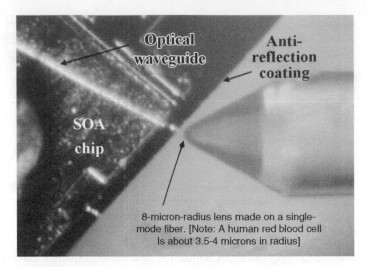

**FIGURE 7.12**   Photograph showing fiber coupling into a semiconductor optical amplifier.

Wires coming up in the center provide electric current for the power required for amplification. The current spreads across the chip into the waveguide. Light passing through the gain material in the waveguide is amplified, generally by as much as 1000 times. Thus, 1 μW in becomes 1 mW out. If too much power is entered, it will damage the output side of the waveguide, unless specifically designed with tapers.

In the case of an SOA, laser light must be channeled into and out of the active layer waveguide as input and output, respectively, to the amplifier. This can be accomplished with a lens or by machining a fiber to the small size of the waveguide, small because the refractive index of III–V materials is around 3.5. Figure 7.12 shows the coupling with fiber. The fiber core, although only 8 μm in diameter, should be pointed and finished with an antireflection coating. Because of the high amplification and nonlinear nature of saturation, many nonlinear processes can be performed with low-power inputs, clipping, frequency changing, and cleaning up pulses [93]. Analysis and modeling of semiconductor optical amplifiers is described in Ref. [25] and some of our research in Refs [91, 92, 94].

# CHAPTER 8

# POWER LASERS

Powerful laser light from 1 W to many gigawatts can be obtained from solids (Nd:YAG-doped glass), liquids (dye lasers) or gases (carbon dioxide, oxygen–iodine). Semiconductor lasers (Chapter 7), can be used for a 1 kW laser (or higher power) by merging outputs for an array of laser diode bars in a way that preserves beam quality (Figure 7.10). A high-power solid-state laser is described in Chapter 13 for allowing nuclear weapon development in an age of test-ban treaties. The free-electron laser, rapidly maturing for very high power from kilowatts to gigawatts, does not depend on bound electron states and is described in Section 11.1.

In discussing power, we must separate continuous wave lasers that remain on for over a second from pulse lasers that are on for only a small part of a second. Pulsed lasers are characterized by pulse energy: peak power times pulse duration. Both energy and duration affect the nature of the impact on the target, the average power for driving the laser, the weight, and the cost. Methods of creating pulsed lasers from continuous ones are described in Chapter 9. High power is often attained by following the laser with a series of optical amplifiers, which are often identical to the initiating laser except for the absence of a resonating cavity (mirrors removed). In this case, a pulse shaper ensures high temporal coherence and spatial filters (a pinhole) after each amplifier ensure high spatial coherence for good beam quality at the final output (Chapter 3). In Section 12.2.2, we describe adaptive optics methods for improving spatial coherence.

In Section 8.1 of this book, we discuss characteristics of high-power lasers. In Section 8.2, we consider solid-state lasers and frequency doubling of solid-state lasers

*Military Laser Technology for Defense: Technology for Revolutionizing 21st Century Warfare,*
First Edition. By Alastair D. McAulay.
© 2011 John Wiley & Sons, Inc. Published 2011 by John Wiley & Sons, Inc.

for visible green light. In Section 8.3, we describe gas dynamic principles, gas dynamic carbon dioxide power lasers used for shooting down cruise and sidewinder missiles in 1983, and chemical oxygen–iodine lasers (COIL) used in the Airborne Laser (ABL) to target intercontinental ballistic missiles (ICBMs) and in the Advanced Tactical Laser (ATL) to target armored vehicles.

## 8.1 CHARACTERISTICS

Power lasers can be categorized by wavelength, power, type, material (gas, solid state, liquid), fiber, semiconductor, pumping technology, continuous wave, or pulse length. Selecting a laser for an application must also include weight, cost, beam quality, and efficiency.

### 8.1.1 Wavelength

Wavelength is one of the most important characteristics. For bound electron lasers, wavelength depends on the material bandgap; only a few easy-to-work-with materials will produce high power economically. According to diffraction (Chapter 3), light of a certain wavelength is relatively unaffected by nonabsorbing particles having a size less than the wavelength while objects larger than the wavelength will be scattered or absorbed. This effect can be observed when waves strike large versus small rocks close to a beach. For some substances where the wavelength is similar to the molecule size, the particle will resonate and absorb the wave energy, for example, water vapor in the atmosphere (Chapter 15). Some applications require visible light, target designators, and spotter lasers, while others prefer invisible infrared (IR) light, range finders, and heat damage lasers.

Sometimes we prefer an eye-safe laser to protect our troops: lasers with emission wavelengths longer than 1.4 $\mu$m are often called *eye safe* because light in this wavelength range is strongly absorbed in the eye's cornea and lens and therefore cannot reach the significantly more sensitive retina. This makes erbium lasers and erbium-doped fiber amplifiers used in 1.5 $\mu$m telecommunication systems less dangerous than Nd:YAG 1 $\mu$m lasers with similar output powers. At longer wavelengths, a $CO_2$ laser at 10.4 $\mu$m, the cornea absorption depth is small, energy is concentrated in a small volume along the surface, and the cornea surface is damaged, apparently a painful injury. Of course, the peak power and energy in a light pulse reaching the eye are also critical. Extensive safety documents and regulations apply to lasers.

### 8.1.2 Beam quality

For a laser beam to deliver energy efficiently to a distant target it must have good beam quality, which means high temporal and spatial coherence and suitable beam convergence (depends on geometry) (Chapter 3). Damaging vehicles with lasers requires high-power lasers similar to those used in manufacturing, where the ability to focus a Gaussian beam to a small spot for cutting and welding has led to a specific quality measurement, spot size times convergence angle (see left side of Figure 8).

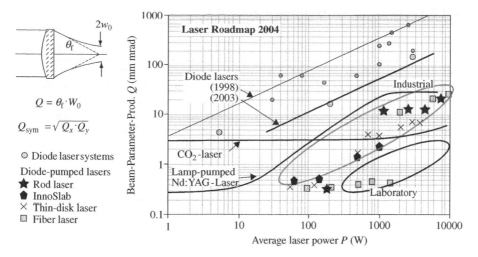

**FIGURE 8.1**  Quality for different types of laser.

Figure 8.1 shows how different types of lasers up to 10 kW power compare in quality [6]. At the top of Figure 8.1, laser diodes arrays are robust, small, and efficient, but have the lowest beam quality (Chapter 7). However, laser diode quality improved by an order of magnitude from 1998 to 2003. Lasers used at very high power, the $CO_2$ laser (Section 8.3.1) that propagates well through the atmosphere, have a higher quality than the solid-state Nd:YAG laser (Section 8.2). Fiber lasers, generally at telecommunication wavelengths, 1.5 µm, are eye safe and have similar quality. They can go to high powers because the fiber lasers may be meters in length and can be placed in a water jacket. Large mode area fibers can carry high power for lasers and power delivery [14, 118]. Oxygen–iodine lasers (Section 8.3.2) were not considered for manufacturing in the past because of the need to store dangerous chemicals. Although this is also a major problem for the military, the need for large power in an airborne platform makes chemical lasers attractive because chemicals provide efficient lightweight energy storage (as in gasoline).

## 8.1.3   Power

Matching the power or peak power to a military application is critical. However, cost, weight, power requirements, complexity, robustness, power requirements, safety, and efficiency are also important. Table 8.1, augmented from Ref. [36], provides an approximate comparison for key properties of the power lasers considered for military applications.

Notes for Table 8.1 by item number.

1. Nd:YAG is a widely used solid-state laser (Section 8.2).
2. The wavelength in item 1 is converted from IR to visible by doubling the frequency (Section 8.2.2).

**TABLE 8.1   Power Lasers Approximate Properties**

| Type | Wavelength | Peak Power (W) | Pulse Length | Repetition Rate | Efficiency |
|------|-----------|----------------|--------------|-----------------|------------|
| 1. Nd:YAG | 1.064 μm | $10^6$–$10^{12}$ | 10 ps–100 ns | 1–100 Hz | $10^{-3}$ |
| 2. Doubled | 532 μm | $10^6$–$10^{12}$ | 10 ps–100 ns | 1–100 Hz | $< 10^{-3}$ |
| 3. Extreme | 351 μm | $0.5 \times 10^{15}$ | ps–ns | 0.001 Hz | $< 10^{-3}$ |
| 4. $CO_2$ | 10.4 μm | $10^8$ | 10 ns–1 μs | 100–500 Hz | $10^{-1}$ |
| 5. Iodine | 1.315 μm | $10^9$–$10^{12}$ | 160 ps–50 ns | 0.014 Hz | $10^{-2}$ |
| 6. KrF | 249 nm | $10^6$–$10^{10}$ | 30 ns–100 ns | 1–100 Hz | $10^{-2}$ |
| 7. LD array | 0.475–1.6 μm | $10^8$ | $10^5$ ps | $10^5$ | 0.5 |
| 8. FEL | $10^{-6}$–2 mm | $10^5$ | 10–30 ns | 0.1–1 Hz | $10^{-2}$ |
| 9. Fiber | 1.018, 1.5 μm | $10^4$ | $10^{-15}$–$10^{-8}$ | $10^{11}$ | 0.03 |

3. Extreme refers to the most powerful lasers in development in the world, the prodigious National Ignition Facility (NIF) in the United States (Chapter 13) and the competitive Megajoule Laser in France. The NIF, currently in test phase, focuses 192 powerful Nd:YAG-doped glass lasers, each with 16 amplifiers, onto a single target with wavelength conversion from 1.064 μm (IR) to 351 nm (UV).

4. The $CO_2$ laser in gas dynamic form was used to disable sidewinder and cruise missiles in the Airborne Laser Lab in 1985 (Sections 8.3.1 and 12.1.2).

5. The COIL in gas dynamic form was used in the current ABL program to shoot down nuclear-armed ICBMs entering the atmosphere (Sections 8.3.2 and 12.2).

6. A Krypton fluoride (UV) laser is included in the table because its ability to generate high power has been demonstrated in military applications.

7. LD array refers to laser diode arrays composed of a large number of laser diodes working together while preserving beam quality (Section 7.2.4).

8. The free-electron laser (FEL), an example of a cyclotron-based laser, is most promising for future destructive beams because almost any wavelength can be generated at very high power (Section 11.1).

9. Optical fiber lasers [32] are evolving from optical fiber amplifiers [30]. They can replaces item 1 in some cases and can be frequency doubled as in item 2. Multiple water-cooled fiber lasers may be combined. The power and efficiency refer to the 1.018 μm wavelength. The second wavelength, 1.5 μm, is less efficient but is eye safe.

## 8.1.4   Methods of Pumping

Energy is provided for population inversion for lasers and optical amplifiers (Chapter 7). Population inversion is achieved by several means or their combinations, such as electrical, optical, thermal, and particle acceleration.

*8.1.4.1 Electrical Pumping* Semiconductor lasers are pumped with electrical current (Chapter 7). The electrons from the electric current jump to a higher energy level to produce population inversion. Laser diodes convert electrons to photons very efficiently. High power is achieved by combining hundreds of laser diodes to produce a single beam, but maintaining quality in the beam severely limits total power. Pulsed $CO_2$ lasers may be pumped with electrical discharge (Section 8.3.1).

*8.1.4.2 Optical Pumping* Incoherent light such as from a flashtube can be used to pump a pulsed laser or an optical amplifier. In this case, incoherent light is converted to coherent light for which all the photons are in lockstep. Often lasers are pumped with many lower powered lasers in a cascade to provide very high power.

*8.1.4.3 Chemical Pumping* Heat is a product of many chemical reactions and the thermal energy can be used to excite particles to a higher energy level (Section 8.3.2).

*8.1.4.4 Gas dynamic Pumping* For gas lasers, power can be substantially reduced by flowing the incoming cold gases or gaseous chemicals through the laser and exhausting the hot gases to prevent overheating of the active medium. Population inversion is achieved by heating the gases to high temperature and pressure and allowing them to escape through nozzles at supersonic speeds, low temperature, and low pressure. In this case, thermal energy causes population inversion. This form of energy conversion from thermal is more like the process in a steam engine.

*8.1.4.5 Particle Acceleration* Cyclotron (or synchrotron) pumping passes a stream of electrons at relativistic speed through an evacuated tube with a periodic field in a free-electron laser (Section 11.1). The flow of relativistic electrons, whose mass and speed depend on each other, causes oscillations with density periodicity that provide high and low energy levels. The externally generated periodic field and the energy of the electron gun select frequency.

## 8.1.5 Materials for Use with High-Power Lasers

High-power lasers described in this chapter and Chapters 9 and 10 are powerful enough to damage optical components used at lower power levels. Consequently, special material should be used that can withstand the high-power levels without damage. Materials for high-power lasers in the 1–3 μm range include fusion-cast calcium fluoride ($CaF_2$), multispectral zinc sulfide (ZnS-MS), zinc selenide (ZnSe), sapphire ($Al_2O_3$), and fused silica ($SiO_2$) [35]. All these materials can be obtained in blanks larger than 175 mm diameter. Sapphire and calcium fluoride require more experience to fabricate than other materials, but all these materials can be polished to the minimal beam deviations required. The final choice of material is made based on the requirements of a system and varies based on maximum laser power, operating environment, durability requirements, wavelength range, size required, and budget. In Section 12.2.2, we discuss the materials used for the high-power optical components and windows (zinc selenide) [34] in the Airborne Laser program.

## 8.2  SOLID-STATE LASERS

### 8.2.1  Principles of Solid-State Lasers

We consider neodymium ($Nd^{3+}$), the most important of the rare earth ions used in solid-state lasers for the military. The most common host material is Nd:YAG, providing coherent light at 1.064 μm [67]. In a laser with extreme power, a glass host allows even higher power (Sections 13.1 and 13.2.1). Nd:YAG lasers are often Q-switched for pulses (Section 9.3).

Solid-state lasers often have a rod-shaped solid. Flashlamps of similar length (20 cm) on one side or both sides are located in polished elliptical cavities. The ends of the rod act as mirrors for the cavity. Figure 8.2a shows a cross section of such a laser [67] with two flashlamps on either side of the rod placed in a double elliptical cavity to focus the pump light into the rod efficiently. The laser is cooled by water, perhaps with dopants to suppress unwanted wavelengths. Figure 8.2b shows a photograph of the top part with the rod and lamps that fits into the bottom part with the double cavity and flow cooling channels.

A powerful pulse Nd:YAG solid-state laser with eight amplification stages is shown in Figure 8.3 [67]. The pulses have energy 750 mJ at 1.064 μm and 40 Hz repetition interval, and the laser is described in detail and analyzed in Ref. [67]. The master oscillator laser is Q-switched to provide the correct pulses. A telescope (Section 1.3.4) reduces the beam diameter for the subsequent system. A Faraday

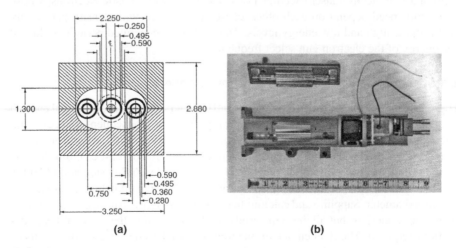

(a)                                                          (b)

**FIGURE 8.2**  Military Nd:YAG laser: (a) cross section showing rod and flashlamps and (b) photograph of upper and lower parts.

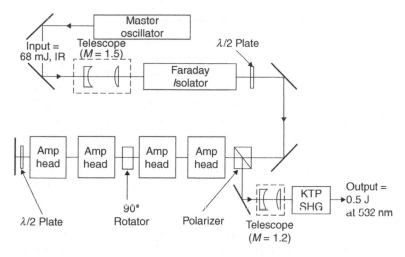

**FIGURE 8.3** Powerful Nd:YAG laser with eight amplification stages.

isolator and a half-wave plate prevent reflections from the subsequent stages from returning to interfere with the master oscillator. The Faraday isolator rotates the polarization clockwise by $\pi/4$ in one direction and anticlockwise by $\pi/4$ on return, resulting in a $\pi/2$ change, so returning light is blocked by a $\pi/2$ polarizer. The light passes through two Nd:YAG amplifiers for which the rods have antireflection coatings to prevent oscillations. Then it passes through a quarter-wave plate (90° phase angle rotator) and two more amplifiers and reflects from a mirror at left. After passing back through the quarter-wave plate, it returns to the polarizing beam splitter with a phase change of 180° and the polarization is rotated by 90° (Section 2.2.1.2). The polarizing beam splitter deflects the new polarization down to the output. A further telescope compresses the beam for high intensity into the nonlinear KTP crystal to double the frequency from 1.064 μm to 532 nm (doubling frequency is equivalent to halving wavelength). Doubling is considered in Section 8.2.2.

Side pumping in this system is performed with laser diode arrays (Section 7.2.4) placed along the rod in place of flashtubes. The efficiency of this system is reported as follows:

1. Laser diode efficiency is 0.35.
2. Conversion of optical pump energy to upper state population is 0.51.
3. Conversion of upper state population is 0.38.
4. Overall efficiency is $0.35 \times 0.51 \times 0.38 = 0.068$ (7%).

In Section 13.2.1, an ultra high-power solid-state laser at the National Ignition Facility is described.

## 8.2.2  Frequency Doubling in Solid State Lasers

High-power (upto 8 kW) infrared light is easily generated with a solid-state Nd:YAG. More than a few milliwatts of infrared laser light is dangerous to the eyes because of the heat it can generate on a very small spot and failure to blink as the eye does not see infrared. Some military applications, such as range finding, target designators, countermeasures, and lidar, use a second harmonic generator to convert the IR to visible green light. Nonmilitary applications include industrial, medical prostate surgery, and replacement of traffic police radar with lidar to reduce incidence of police cancer.

We derive equations for second harmonic generation (SHG) that doubles the laser frequency $f$, or equivalently, as $\lambda = c/f$, halves the wavelength $\lambda$. In this case, for a solid-state Nd:YAG laser, the IR wavelength $\lambda = 1.064 \, \mu m$ is converted to green at $\lambda = 0.5 \, \mu m$ [67, 176]. A nonlinear crystal performs SHG.

### 8.2.2.1  Nonlinear Crystal for SHG
Doubling is accomplished by passing the IR light into a nonlinear crystal [176]. In elastic media, the electrons are pulled back and forth by the alternating electric field of the IR light. The polarization of the crystal medium represents an induced electric field that opposes the IR field influence and can be detected by suitably arranging a system. Centrosymmetric crystals [176] such as sodium chloride (common salt), NaCl, cannot be used for SHG because they have structural symmetry about a sodium ion ($Na^+$), shown as dots in Figure 8.4a, which causes them to polarize linearly and equally for plus and minus electric fields (Figure 8.4b). Hence, there is no change in frequency between incident light and induced polarization.

In noncentrosymmetric crystals, such as potassium dihydrogen phosphate (KDP), or zinc sulfide (Figure 8.5a), the central sulfur ion does not see symmetry between NE and SW zinc ions because the zinc in this case is at different depths in these two places. Consequently, there is a nonlinear relation between incident electric field and polarization (Figure 8.5b). Electrons are more easily pulled to one side than the other. An incident field is suppressed more in the positive polarization in the crystal than in the negative polarization. When decomposed into Fourier components, the distorted polarization sine wave shows presence of double frequencies. Nonlinear polarization

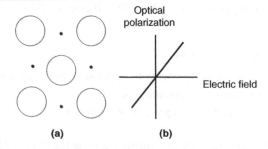

**(a)**                          **(b)**

**FIGURE 8.4**  Symmetric crystal: (a) structure and (b) optical polarization.

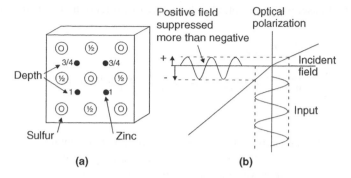

**FIGURE 8.5**   Noncentrosymmetric crystal: (a) structure and (b) optical polarization.

$(P_i)$ can be described by a Taylor series

$$P_i = E_0 \chi_{ij} E_j + 2d_{ijk} E_j E_k + 4\chi_{ijkl} E_j E_k E_l + \cdots \qquad (8.1)$$

where $\chi_{ij}$ is the first-order or linear susceptibility, $d_{ijk}$ is the second-order nonlinear susceptibility (allows SHG), and $\chi_{ijkl}$ is the third-order nonlinear susceptibility (for different nonlinear effects). We are interested in the second-order nonlinear effects for SHG for which the polarization depends on the product of two electric fields that can be in different directions ($j$ and $k$). In particular, from the second term on the right-hand side of equation (8.1), we desire the polarization $P_i^{2\omega}$, in the $i$th direction at double frequency $2\omega$, resulting from the interaction of two incident IR electric fields, $E_j^{\omega}$ and $E_k^{\omega}$, in the $j$ and $k$ orthogonal directions to be

$$P_i = 2d_{ijk} E_j E_k \qquad (8.2)$$

We convert terms $P$ and $E$ to exponentials using

$$E = \mathrm{Re}\,\{E_0\} = \frac{1}{2}\left\{ E_0 e^{i\omega t} + E_0 e^{-i\omega t} \right\} = \frac{1}{2}\left\{ E_0 e^{i\omega t} + \mathrm{c.c.} \right\} \qquad (8.3)$$

where c.c. is complex conjugate and subscript 0 denotes a phasor. (Note that here we used the physics nomenclature time harmonic $\exp\{i\omega t\}$ and elsewhere we have used the Re $\exp\{-j\omega t\}$.) Using equation (8.3), we have several ways to produce second harmonic ($2\omega$) frequencies described in equation (8.2). Two ways are representative: select the two $E_0$ fields so that their frequencies $\omega_1$ and $\omega_2$ during nonlinear mixing will produce terms in $2\omega_1$ or select the two $E_0$ fields with the same frequency $\omega$, which is half the final double frequency $2\omega$.

For the first case, with addition of fields at $\omega_1$ in $j$ direction and $\omega_2$ in $k$ direction

$$\frac{1}{2} P_{0i}^{\omega_1+\omega_2} e^{i(\omega_1+\omega_2)t} + \text{c.c.}$$

$$= 2d_{ijk} \frac{1}{2} \left\{ E_{0j}^{\omega_1} e^{i\omega_1 t} + E_{0j}^{\omega_2} e^{i\omega_2 t} + \text{c.c.} \right\} \frac{1}{2} \left\{ E_{0k}^{\omega_1} e^{i\omega_1 t} + E_{0k}^{\omega_2} e^{i\omega_2 t} + \text{c.c.} \right\}$$

$$\tag{8.4}$$

From equation (8.4), we extract those terms of the mixing that contain $\omega_1 + \omega_2$ (which add to become the double frequency $2\omega_1$). Note that only the two cross terms give $\omega_1 + \omega_2$. At the same time, we cancel the term $e^{i\omega_1 t} e^{i\omega_2 t} = e^{i(\omega_1+\omega_2)t}$ from each side to obtain the phasor equivalent of equation (8.2):

$$P_{0i}^{\omega_1+\omega_2} = 2d_{ijk} E_{0j}^{\omega_1} E_{0k}^{\omega_2} \tag{8.5}$$

For the second case, we do not desire the sum of two different frequencies $\omega_1$ and $\omega_2$ to equal $2\omega$, $(\omega_1 + \omega_2 = 2\omega_1)$, but rather select both frequencies to be $\omega$ or $\omega_1 = \omega_2 = \omega$ in equation (8.4). Once again select only terms at the second harmonic frequency $2\omega$. Using the fact that the two brackets on the right-hand side of equation (8.4) equal each other gives us additional 2 factor (relative to the first case) and canceling the exponents gives

$$P_{0i}^{2\omega} = d_{ijk} E_{0j}^{\omega} E_{0k}^{\omega} \tag{8.6}$$

As there are three rectangular coordinate directions $(x, y, z)$ and six combinations of pairs of fields from three directions, the second-order nonlinear coefficient $d_{ijh}$ is a $3 \times 6$ tensor (compressed from $3 \times 9$ by merging different index orders):

$$\begin{bmatrix} P_{0x}^{2\omega} \\ P_{0y}^{2\omega} \\ P_{0z}^{2\omega} \end{bmatrix} = \begin{bmatrix} d_{11} d_{12} d_{13} d_{14} d_{15} d_{16} \\ d_{21} d_{22} d_{23} d_{24} d_{25} d_{26} \\ d_{31} d_{32} d_{33} d_{34} d_{35} d_{36} \end{bmatrix} \begin{bmatrix} E_{0x} E_{0x} \\ E_{0y} E_{0y} \\ E_{0z} E_{0z} \\ 2E_{0y} E_{0z} \\ 2E_{0z} E_{0x} \\ 2E_{0x} E_{0y} \end{bmatrix} \tag{8.7}$$

In most crystals, only a few of the coefficients in the matrix are nonzero. For KDP, only $d_{14}$, $d_{25}$, and $d_{36}$ have significant values, giving

$$P_x^{2\omega} = 2d_{14} E_{0y}^{\omega} E_{0z}^{\omega}$$

$$P_y^{2\omega} = 2d_{25} E_{0x}^{\omega} E_{0z}^{\omega}$$

$$P_z^{2\omega} = 2d_{36} E_{0x}^{\omega} E_{0y}^{\omega} \tag{8.8}$$

Note that for $E_z = 0$, the last equation of equation (8.8) represents an incident TEM mode in the $z$ direction, causing second harmonic in the $z$ direction. The angle of incidence of the incident wave at frequency $\omega$ relative to the crystal axes must be selected carefully for phase matching between the input wave at frequency $\omega$ and the output second harmonic or frequency-doubled wave at frequency $2\omega$ to achieve efficient SHG. Second-order nonlinear coefficients are provided in Ref. [176].

### *8.2.2.2   Electromagnetic Wave Formulation of SHG*   We follow the approach of Ref. [176] and use labels $E_1$, $E_2$, and $E_3$ for phasor fields $x$, $y$, and $z$, respectively, for a wave propagating in the $z$ direction. This $z$ is not the same as the earlier used $z$ axis of the crystal. Consider only the second case in Section 8.2.2.1, equation (8.6), there are two coupled waves, at frequencies $\omega$ and $2\omega$, propagating concurrently through the crystal in wave propagation direction $z$ during SHG. The field $E_1$ from the inputs at frequency $\omega_1 = \omega$ decays along the crystal according to $dE_1/dz < 0$ as power is converted to the second harmonic at $2\omega$. The desired wave $E_3$ at frequency $\omega_3 = 2\omega$ (double frequency) increases from zero along the crystal as its power builds up according to $dE_3/dz > 0$. Separating out polarization of each field $E_{0i}^{\omega_1} = a_{1i}E_1$ and $E_{0i}^{\omega_3} = a_{3i}E_3$, we write the instantaneous (a function of $z$ and $t$) nonlinear polarization for the $\omega$ and $2\omega$ waves, using equations (8.5) and (8.6), and $\omega_3 - \omega_1 = 2\omega_1 - \omega_1 = \omega_1$, as [176]

$$\left[P_{NL}^{\omega_3-\omega_1}(z,t)\right]_i = d_{ijk}a_{3j}a_{1k}E_3E_1^*e^{i[(\omega_3-\omega_1)t-(k_3-k_1)z]} + \text{c.c.}$$

$$\left[P_{NL}^{2\omega_1}(z,t)\right]_i = \frac{1}{2}d_{ijk}a_{1j}a_{1k}E_1E_1e^{i(2\omega_1t-2k_1z)} + \text{c.c.} \qquad (8.9)$$

where summations over repeated indices are assigned $d = \sum_{ijk} d_{ijk}a_{1i}a_{2j}a_{3k}$. Following a standard procedure, we substitute in turn the polarizations from equation (8.9) into the wave equation. The wave equation, assuming materials are not conductive (conductivity $\sigma = 0$), is

$$\nabla^2 E_1(z,t) = \mu_0 \frac{\partial^2}{\partial t^2}(\epsilon_1 E_1(z,t) + P_{NL}) = \mu_0\epsilon_1\frac{\partial^2 E_1(z,t)}{\partial t^2} + \mu_0\frac{\partial^2}{\partial t^2}P_{NL} \quad (8.10)$$

Substituting the first equation of equation (8.9) into equation (8.10) gives

$$\nabla^2 E_1^{\omega_1}(z,t) = \mu_0\epsilon_1\frac{\partial^2 E_1^{\omega_1}(z,t)}{\partial t^2} + \mu_0 d\frac{\partial^2}{\partial t^2}\left\{E_3E_1^*e^{i[(\omega_3-\omega_1)t-(k_3-k_1)z]} + \text{c.c.}\right\}$$

$$(8.11)$$

As $E_1$ in equation (8.11) is dependent on the product of $z$-dependent terms, the left-hand side can be written in terms of phasor $E_1$ as

$$\nabla^2 E_1^{(\omega_1)}(z, t) = \frac{1}{2} \frac{\partial^2}{\partial z^2} \left[ E_1(z) e^{i(\omega_1 t - k_1 z)} + \text{c.c.} \right]$$

$$= -\frac{1}{2} \left[ k_1^2 E_1(z) + 2ik_1 \frac{dE_1}{dz}(z) \right] e^{i(\omega_1 t - k_1 z)} + \text{c.c.} \quad (8.12)$$

where we assumed the slowly varying amplitude approximation to eliminate $d^2 E_1(z)/(dz^2)$ with (see Section 2.1.1)

$$\frac{d^2 E_1(z)}{dz^2} \le k_1 \frac{\partial E_1}{\partial z}(z) \quad (8.13)$$

Substituting equation (8.12) for the left-hand side of equation (8.11) and using $\partial^2/\partial t^2 \equiv -\omega_1^2$,

$$-\frac{1}{2} \left[ k_1^2 E_1(z) + 2ik_1 \frac{dE_1}{dz}(z) \right] e^{i(\omega_1 t - k_1 z)} + \text{c.c.}$$

$$= -\mu_0 \epsilon_1 \omega_1^2 \left[ \frac{E_1(z)}{2} e^{i(\omega_1 t - k_1 z)} \right] + \text{c.c.}$$

$$-\mu_0 d \omega_1^2 \left[ E_3 E_1^* e^{i\{(\omega_3 - \omega_1)t - (k_3 - k_1)z\}} \right] + \text{c.c.} \quad (8.14)$$

As $k_1^2 = \mu_0 \epsilon_1 \omega_1^2$, the terms like this on both sides cancel. Also, as $\omega_3 = 2\omega$ and $\omega_1 = \omega$, $\omega_3 - \omega_1 = \omega$, the term $e^{i(\omega_1 t)}$ at left cancels with $e^{i(\omega_3 - \omega_1)t}$ at right:

$$-ik \frac{dE_1}{dz} e^{-i(k_1 z)} + \text{c.c.} = -\mu_0 d \omega_1^2 \left[ E_3 E_1^* e^{-i(k_3 - k_1)z} + \text{c.c.} \right] \quad (8.15)$$

or the rate of decay of the incident field at frequency $\omega$ with distance $z$ along the crystal

$$\frac{dE_1}{dz} = -i\omega_1 \sqrt{\frac{\mu_0}{\epsilon_1}} dE_3 E_1^* e^{-i(k_3 - 2k_1)z} \quad (8.16)$$

where we used $(\mu_0 \omega_1^2)/(ik) \equiv -i\omega_1 \sqrt{\mu_0/\epsilon_1}$ with $k = \omega \sqrt{\mu_0 \epsilon_1}$.

Similarly, substituting the second equation of equation (8.9) into the wave equation (8.10) gives the rate of increase of the second harmonic at $2\omega$ with distance $z$ along the crystal: ($\omega_3 = 2\omega$)

$$\frac{dE_3}{dz} = -\frac{i\omega_3}{2} \sqrt{\frac{\mu_0}{\epsilon_3}} dE_1 E_1^* e^{i(k_3 - 2k_1)z} \quad (8.17)$$

Equations (8.16) and (8.17) describe how the incident waves at frequency $\omega$ decay and the wave at frequency $2\omega$ grows. It can be shown that in the absence of loss, the power of combined waves is constant; that is, the power lost with distance in the input waves at frequency $\omega$ equals the growth in power with distance in the $2\omega$ wave [176]:

$$\frac{d}{dz}(\sqrt{\epsilon_1}|E_1|^2 + \sqrt{\epsilon_3}|E_3|^2) = 0 \qquad (8.18)$$

For high-intensity lasers and phase matching by angle adjustment, conversion efficiency of 100% is possible.

### 8.2.2.3  *Maximizing Second Harmonic Power Out*  Following Ref. [176], phase matching for the $\omega$ and $2\omega$ waves over sufficient distance enhances efficiency. Its effect can be seen by initially assuming that negligible power is lost for $\omega$, which occurs after only a short distance into the crystal or for lower intensity beams. In this case, $E_1$ is constant and only equation (8.17) showing the growth of second harmonic ($2\omega$) from zero needs be considered:

$$\frac{d}{dz}E_3^{(2\omega)} = -i\omega\sqrt{\frac{\mu_0}{\epsilon_3}}d|E_1|^2 e^{i\Delta kz} \qquad (8.19)$$

where we defined $\Delta k = k_3 - 2k_1$ and $\omega = \omega_3/2$. The field of the second harmonic $E_3$ is obtained by integration along the crystal length $L$:

$$E_3^{(2\omega)}(L) \equiv \int_{z=0}^{L} \frac{d}{dz}E_3^{(2\omega)}dz = \frac{-i\omega}{n^2}\sqrt{\frac{\mu_0}{\epsilon_0}}d|E_1|^2\left(\frac{e^{i\Delta kL} - 1}{i\Delta k}\right) \qquad (8.20)$$

where we used $\epsilon_3 = n^2\epsilon_0$. Intensity of the second harmonic generated is

$$I^{(2\omega)}(L) = E_3^{2\omega}(L)E_3^{*(2\omega)}(L) = \frac{\omega^2 d^2}{n^2}\left(\frac{\mu_0}{\epsilon_0}\right)L^2|E_1|^2\frac{\sin^2(\Delta kL/2)}{(\Delta kL/2)^2} \qquad (8.21)$$

where we used $\exp\{i\Delta kL\} - 1 = \exp\{i\Delta kL/2\}(\exp\{i\Delta kL/2\} - \exp\{-i\Delta kL/2\}) = \exp\{i\Delta kL/2\}(-2i\sin(\Delta kL/2))$ and multiplied top and bottom by $L^2$. From intensity of positive-going $\omega$ wave and intrinsic impedance $\eta = \sqrt{\mu_0/\epsilon_0}$,

$$I^{(\omega)} = \frac{1}{2}n\sqrt{\frac{\epsilon_0}{\mu_0}}|E_1|^2 \quad \text{or} \quad |E_1|^2 = 2I^{(\omega)}\sqrt{\frac{\mu_0}{\epsilon_0}}\frac{1}{n} \qquad (8.22)$$

Dividing equation (8.21) by equation (8.22) gives the efficiency of second harmonic generation:

$$
\eta_{SHG} = \frac{I^{(2\omega)}}{I^{(\omega)}} = \frac{w^2 d^2}{n^2} \frac{\mu_0}{\epsilon_0} L^2 \left( 2I^{(\omega)} \sqrt{\frac{\mu_0}{\epsilon_0}} \frac{1}{n} \right) \frac{\sin^2 \Delta k L/2}{(\Delta k L/2)^2}
$$

$$
= \frac{2\omega^2 d^2}{n^3} \left( \frac{\mu_0}{\epsilon_0} \right)^{3/2} L^2 \frac{\sin^2 \Delta k L/2}{(\Delta k L/2)^2} I^{\omega} \tag{8.23}
$$

Thus, the efficiency of second harmonic generation is proportional to the intensity of the input beam. So for maximizing the intensity of the second harmonic, we should select material with a high value for $d$ (KDP is good), use a longer length $L$, use a large intensity input, and have the $\omega$ wave propagate at the same propagation constant (wave number) $k$ as the $2\omega$ wave. The last condition gives $\Delta k = k^{(2\omega)} - k^{(\omega)} = 0$ for which the phases of the two waves are matched. Unfortunately, waves copropagating through a crystal at frequencies $\omega$ and $2\omega$ do not have the same propagation constant, so $\Delta k$ cannot equal zero. The next section addresses this issue.

**8.2.2.4  Phase Matching for SHG**  Phase matching to achieve $\Delta k = 0$ is accomplished by selecting the angle at which the incident wave at frequency $\omega$ strikes the crystal. A uniaxial anisotropic crystal such as KDP exhibits birefringence as shown in its index ellipsoid (Figure 8.6). The equation for the index ellipsoid is

$$
\frac{x^2}{n_o} + \frac{y^2}{n_o} + \frac{z^2}{n_e} = 1 \tag{8.24}
$$

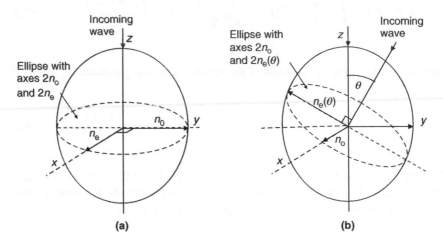

**FIGURE 8.6**  Process for phase matching for SHG: (a) index ellipsoid and (b) finding angle $\theta$ for phase matching.

An incoming wave sees refractive indices $n_o$ for its $x$ and $y$ polarized components and $n_e$ for its $z$ polarized one. In a uniaxial crystal, subscripts $n_o$ and $n_e$ refer to ordinary and extraordinary directions, respectively, the latter being extraordinary because its refractive index differs from the other two orthogonal directions that have the ordinary refractive index $n_o$. If the incoming wave propagating in the $z$–$y$ plane is tilted at an angle $\theta$ with respect to the $z$ axis (Figure 8.6b), the wave strikes an elongated ellipse (marked by dashes) at normal incidence. The refractive index $n_o$ is unchanged for the perpendicular component of the incoming wave, while the refractive index for the parallel component is stretched from $n_e$ to $n_e(\theta)$. The stretch is computed as follows. For the ellipse including the $z$, $y$, and incident wave directions,

$$\frac{y^2}{n_o^2} + \frac{z^2}{n_e^2} - 1 \tag{8.25}$$

Dropping a perpendicular from $n_e(\theta)$ to the $y$ axis gives

$$y = n_e(\theta) \cos \theta \tag{8.26}$$

and to the $z$ axis gives

$$z = n_e(\theta) \sin \theta \tag{8.27}$$

Substituting equations (8.26) and (8.27) into equation (8.25) gives an equation for determining $n_e(\theta)$ in order to achieve phase matching with a specific crystal:

$$\frac{1}{n_e^2(\theta)} = \frac{\cos^2 \theta}{n_o^2} + \frac{\sin^2 \theta}{n_e^2} \tag{8.28}$$

We write equation (8.28) for the second harmonic wave at frequency $2\omega$ as

$$\frac{1}{\left[n_e^{(2\omega)}\right]^2} = \frac{\cos^2 \theta}{\left[n_o^{(2\omega)}\right]^2} + \frac{\sin^2 \theta}{\left[n_e^{(2\omega)}\right]^2} \tag{8.29}$$

For phase matching of $\omega$ and $2\omega$ waves, we require

$$n_e^{2\omega}(\theta) = n_o^{(\omega)} \tag{8.30}$$

so that velocities and propagation constants for the two waves are equal, velocity $= c/n_e^{(2\omega)}(\theta) = c/n_o^{(\omega)}$. Substituting equation (8.30) into the left-hand side of equation (8.29) gives an equation for which the optimum angle $\theta$ for phase matching can

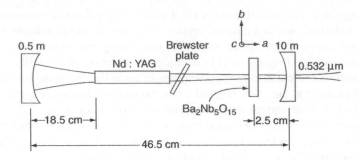

**FIGURE 8.7** Configuration for second harmonic generation, crystal inside laser cavity.

be computed as

$$\frac{1}{\left[n_o^{(\omega)}\right]^2} = \frac{\cos^2 \theta}{\left[n_o^{(2\omega)}\right]^2} + \frac{\sin^2 \theta}{\left[n_e^{(2\omega)}\right]^2} \tag{8.31}$$

Solving for $\theta$ by relating $\cos^2 \theta = 1 - \sin^2 \theta$ gives

$$\sin^2 \theta = \frac{\left[n_o^{(\omega)}\right]^{-2} - \left[n_o^{(2\omega)}\right]^{-2}}{\left[n_e^{(2\omega)}\right]^{-2} - \left[n_o^{(2\omega)}\right]^{-2}} \tag{8.32}$$

Figure 8.7 shows a simple configuration for SHG [67] that allows a large TEM$_{00}$ mode volume in a crystalline Nd:YAG rod and high intensity in a nonlinear noncentrosymmetric barium sodium niobate crystal. The laser cavity for the Nd:YAG laser is formed by two concave mirrors that highly reflect the YAG infrared wavelength of 1.064 μm. The right-hand side mirror is transparent to the half-wavelength green 532 μm (double frequency) generated in the barium sodium niobate crystal. The latter is placed at the location of maximum intensity in the cavity. Green light that travels left from this crystal is absorbed by the Nd:YAG rod. KDP, potassium dideuterium phosphate (KH$_2$PO$_4$), is commonly used for power applications because it has a high damage threshold and high optical quality [67].

## 8.3  POWERFUL GAS LASERS

We consider gas dynamic carbon dioxide (Section 8.3.1) and chemical oxygen–iodine lasers (Section 8.3.2).

## 8.3.1 Gas Dynamic Carbon Dioxide Power Lasers

The best source for near-monochromatic, very high-power lasers is the gas dynamic laser, particularly for continuous wave lasers [119]. Although solid-state lasers have higher molecular number densities than gases with potentially higher gains, they have substantial drawbacks, especially for continuous wave. The drawbacks result by uneven heating of the solid material from pumping and lasing. The National Ignition Facility laser (Chapter 13) is solid state but produces pulses of less than 1 ns duration and is expected to need 5 h between shots to cool down. In a solid-state laser, even if the high power does not damage the solid medium, it creates distortions in the medium that degrades spatial coherence and hence beam quality. With a gas, medium temperature variations across the gas are less than those in a solid and degradation to the medium occurs only at extreme intensities due to photoionization. However, pumping and lasing generate waste heat that builds up and can limit laser power to kW in a $CO_2$ laser, still enough to burn a hole through quarter-inch steel in a few seconds but not enough for a military damage laser. The heat buildup is eliminated with a gas dynamic laser.

***8.3.1.1 Principles of Gas Dynamic Laser*** Continuous wave power is increased by orders of magnitude in a gas dynamic laser by flowing the gases through the laser; then, cooler gas flows in and hot gas flows out of the laser [36]. This prevents heat buildup in the laser. Figure 8.8 illustrates a gas dynamic laser. In addition to keeping the laser cool, the gas dynamic laser also permits an alternative method of pumping a gas laser that is not available to non-gas lasers [119]. A mixture of gases for lasing is heated in a high-pressure combustion chamber (plenum) to thousands of degrees and pressures of around $10^6$ Pa. The international standard (SI) unit of pressure since 1954 is the pascal (Pa); pounds per square inch (psi) is defined as 1 psi $= 6.894 \times 10^{-3}$ Pa, and 1 atm $= 101,325$ Pa at sea level.

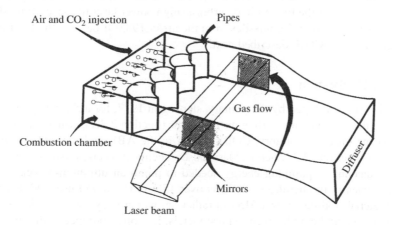

**FIGURE 8.8** Structure of a gas dynamic laser.

The population densities are satisfied by Boltzmann's equation (Section 7.1). The gas is suddenly allowed to leave through an array of nozzles, aerodynamically shaped for smooth flow and with opening holes of typically 1–2 mm. A supersonic or hypersonic expansion of the gas occurs: supersonic means moving at up to four times the speed of sound in air, which is 343 m/s and hypersonic means faster than four times the speed of sound or Mach 4. The temperature and pressure fall by an order of magnitude due to this expansion, which results in population inversion. In the case of a $CO_2$ gas dynamic laser, the population inversion is due to the nonequilibrium flow, allowing rapid decay of the energy in the lower temperature and lower energy level, and slow decay of the vibrational $N_2$ energy that allows excitation transfer to the $CO_2$ [119]. The population inversion is due to the temperature differences or represents thermal pumping. Thus, thermal energy is converted to coherent laser light. In contrast, thermal energy is converted to mechanical energy in machines using a Carnot cycle.

### 8.3.1.2    Using a Gas Dynamic Laser to Shoot Down Missiles    A mixture of carbon dioxide (14%), nitrogen (85%), and water (1%) for laser pumping was generated by burning together carbon monoxide (CO) for fuel and gaseous nitrous oxide ($N_2O$) as an oxidizer at a temperature of 3000 K and pressure of $5.5 \times 10^6$ Pa. The hot gas flows through a horizontal 2 m long distribution manifold [34] (Figure 8.9) and then turns vertical and moves through a band of nozzles into a horizontal laser cavity with mirrors at each end. The nozzles have a space about $1.6 \, mm^2$ for the gas to expand through. The nozzles are shaped for aerodynamic flow to provide good beam quality. The nozzles accelerate the flow by six times the speed of sound, Mach 6, to create a population inversion for lasing capable of producing $10^{25}$ photons. The spent gas stream moves up through the diffuser, which brings its pressure back to atmospheric level for venting out of the aircraft.

The $CO_2$ gas dynamic laser was mounted upside down in the Airborne Laser Laboratory (ALL) on an NKC-135 airplane [34], as shown in Figure 8.10, together with its fuel supply system (FSS) and optical steered output. The laser system in Figure 8.10 [34] was the first military airborne high power laser to successfully shoot down sidewinder and cruise missiles from the air in 1983. It became the forerunner to the gas dynamic COIL described next.

### 8.3.2    COIL System

The COIL was developed by the military, starting in 1977, for airborne applications to replace the gas dynamic carbon dioxide laser (Section 8.3.1), which was used in the Airborne Laser Laboratory. COIL is used in the ABL and the ATL programs described in Section 12.2. Chemical lasers use a chemical reaction to pump the laser; the large amount of pumping energy needed to pump an ultrahigh-power laser is carried efficiently in chemicals. COIL is over 15% efficient and uses iodine as the lasing material. Iodine emits 1.315 μm radiation, which is eye safe, carries well in an optical fiber, and couples well to most metals. It is often pumped with ultraviolet light, which breaks iodine molecules down by photolysis, leaving an energized iodine

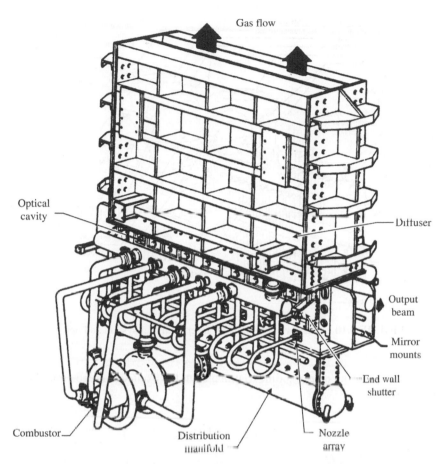

**FIGURE 8.9** Physical structure of the ALL gas dynamic $CO_2$ laser that shot down a missile in 1983.

atom for population inversion. In the case of COIL, chemically pumping iodine to an excited state is performed by excitation transfer pumping that involves two stages. In the first stage, oxygen is excited to a metastable state known as singlet delta oxygen, $O_2^*(1\Delta)$:.

$$O_2 \longrightarrow O_2^*(1\Delta) \qquad (8.33)$$

Then, in the second stage, the singlet delta oxygen is merged with iodine to transfer the excitation from the oxygen to the iodine molecule for population inversion of the iodine:.

$$O_2^* + I \longrightarrow O_2 + I^* \qquad (8.34)$$

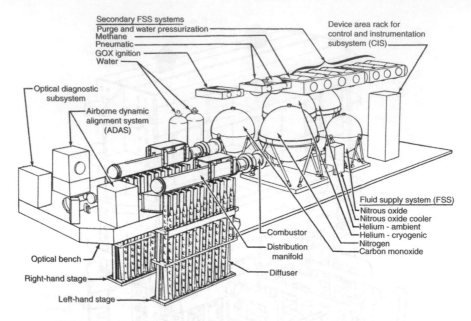

**FIGURE 8.10**  ALL gas dynamic $CO_2$ laser with fuel system and optics.

After pumping, lasing causes iodine to move from the excited stage to emit photons at 1.315 μm:

$$I^2P_{1/2} \longrightarrow I^2P_{3/2} \tag{8.35}$$

We consider the two pumping stages separately: singlet delta oxygen production, equation (8.33), and excited iodine production, equation (8.34).

### 8.3.2.1  Singlet Delta Oxygen Production

A lot of methods have been proposed for generation of singlet delta oxygen and many have been tested. The most convenient method for an airborne application, at left-hand side of Figure 8.11 [31], is to mix basic hydrogen peroxide (BHP) with chlorine, in which the word basic refers to the addition of a base such as KOH (or NaOH). The liquid BHP passes through as many as 12,000 orifices to create drops for a high surface area contact with the gas. The chlorine gas passes through an array of orifices and most of the chlorine is used up in the chemical reaction. The chemical reaction produces heat, potassium chloride, and oxygen in an excited state (singlet delta oxygen) with a spontaneous lifetime of 45 min. This operation occurs at left-hand side in the reactor as shown in Figure 8.11 [31] (from U.S. patent 6,072,820). Other methods for exciting (pumping) the oxygen are often simpler or faster but may be less efficient. For example, electric discharge, such as used for a $CO_2$ electric discharge laser EDL, is used in Ref. [146]. Another example claims to provide a laser pulse more quickly, useful in

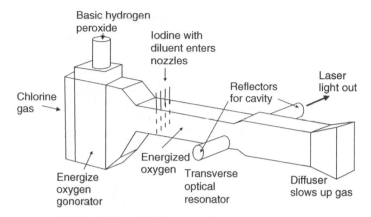

**FIGURE 8.11** Chemical oxygen–iodine laser.

an airborne weapon, by supersonic expansion of oxygen to produce a molecular beam that contains clusters of oxygen and singlet delta oxygen [19].

### 8.3.2.2 Excited Iodine Production for Lasing

The singlet delta oxygen is now mixed with iodine in an array of nozzles in the second stage shown in Figure 8.11 [31]. The singlet delta oxygen passes through interstitial spaces between nozzles that speeds it up, for example, from Mach 0.1 to Mach 1 (one times the speed of sound, which is 343 m/s at atmospheric pressure). The high-pressure iodine gas at, say, 100 psi is mixed with a nitrogen diluent (which does not participate in the chemical reaction) and the mixture is released through the aerodynamic nozzles that speed it up to Mach 5 (five times the speed of sound) and the pressure falls from 100 to 0.2 psi. Turbulent mixing is effective because fast moving iodine merges with slower moving singlet delta oxygen. The resulting output has a speed of Mach 3.5 and temperature of 100 K. This will cool the laser and lowers pressure for optimum gain in the laser. The laser beam is fixed at right angles to the gas flow by setting mirrors to form a cavity in this direction. Following the laser, expansion in a diffuser performs a function opposite to that of nozzles. It recovers the pressure from 0.2 to 3 psi from whence it can be pumped after scrubbing of toxic chemicals or absorbed for a closed system. Note that the nozzles are not used in the same way as in the $CO_2$ laser to produce population inversion because the chemical reaction provides the heat needed for thermal population inversion. However, the nozzles provide the temperature and pressure for optimum laser gain.

### 8.3.2.3 Absorbing the Waste Gases

The hot waste gases can be used to detect the presence of an airplane and recognize that it is carrying a laser weapon. So, to remain undetected and for safety, the chemical oxygen–iodine laser must absorb all the hot nonatmospheric waste gases without exhausting them out of the plane. A cryoabsorption vacuum pump is shown in Figure 8.12 [165] (from U.S. patent 6,154,478). The chemical oxygen–iodine laser is at the left. The waste gases pass

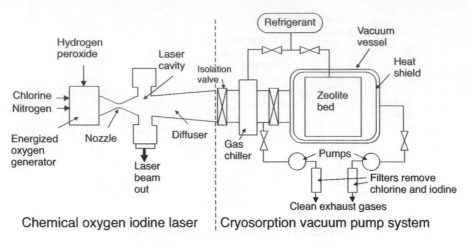

Chemical oxygen iodine laser | Cryosorption vacuum pump system

**FIGURE 8.12**   Absorbing waste gases in COIL.

through a valve into the cryoabsorption vacuum pump in the right box. The gases pass into a chiller at 80 K, cooled by a refrigerant of liquid nitrogen or argon in a Dewar. The chiller condenses or freezes out chlorine gas, iodine, and water vapor, which are trapped on the cooled surfaces. Only cold and dry nitrogen and oxygen are left. These are absorbed by granulite zeolite in a vacuum vessel at 80 K. Absorption of the gases acts as a vacuum pump to draw the gases through the complete system until the zeolite bed fills and the pressure is no longer adequate. The valves are then controlled to regenerate the system and the pump filtered clean gases out from the chiller and zeolite bed at close to atmospheric pressure and temperature.

# CHAPTER 9

# PULSED HIGH PEAK POWER LASERS

Continuous wave lasers can be modified to provide pulsed lasers that are preferable in many situations. A pulse shaping network can be used to form pulses from the output of a continuous wave laser, often used for very high laser powers. In Section 9.1 we discuss situations in which a pulsed laser may be advantageous over a continuous one. The use of two gratings to change pulse duration is discussed. In Section 9.2, we describe how a laser may be mode locked to become a pulsed laser. In Section 9.3, we describe Q-switching to generate pulses. In contrast to mode locking, the laser does not need many modes. In Section 9.4.1, we describe how arrays of lasers can be synchronized to simultaneously focus beams in space and time.

## 9.1 SITUATIONS IN WHICH PULSED LASERS MAY BE PREFERABLE

A pulsed laser is preferable to a continuous wave laser in many applications.

1. *Fast Changing Environments*: Pulsed lasers are effective when the environment and/or geometry is changing rapidly.
2. *Lidar and Ladar*: The time between sending a pulse and receiving its reflection or scattering allows distance estimation or surface profiling for target identification and ground maps. Scanning a pulsed system detects chemical weapons and dangerous weather patterns.

*Military Laser Technology for Defense: Technology for Revolutionizing 21st Century Warfare*, First Edition. By Alastair D. McAulay.
© 2011 John Wiley & Sons, Inc. Published 2011 by John Wiley & Sons, Inc.

3. *Low Power for Mobile Applications*: For aircraft, mobile vehicles, and person-nel, pulse lasers reduce average power, resulting in smaller size and less weight for batteries and generators. For the same peak power, a pulse with 1% duty cycle or a fixed pulse at 1/100 the repetition rate lowers average power by 100. Other possible ways of reducing power include generating it from solar energy or from the recoil when firing conventional weapons.

4. *Adjusting Pulse Length*: Pulse length can be adjusted with pulse shaping net-works to match an application. For example, a pair of parallel gratings can narrow a pulse that has frequencies spread out in time: one spreads the differ-ent frequencies that are then delayed differently while the other brings them back together, having moved the energy for the frequencies closer together in time [72]. For amplifying high-power pulses, it is common to broaden the pulse before amplification to minimize amplifier damage and shorten it again after amplification.

5. *Energy*: The energy in a pulse is its peak power times its duration. Energy, pulse duration, and wavelength must match the application. With usual power lasers for heat damage, longer than nanosecond pulses are used.

6. *Selecting Pulse Length*: For very long pulses, heat damage can spread to neighboring regions, which may match a target application but could also be detrimental, for example, in surgery. Very short pulses (picoseconds) can vaporize materials, bypassing the liquid phase, in laser-induced breakdown spectroscopy [120]. Also, they can produce shock waves that result in metal failure and the plasma is thought to interfere with jet engine intake. Such pulses can break down air like lightning, creating a conducting path through which high-power dangerous microwaves, not normally focusable, can travel.

7. *Light Bullets*: The light from a pulsed laser can look like a light bullet. For example, if the pulse length $\Delta t$ is 1 ns ($10^{-9}$), 1 ps ($10^{-12}$), and 1 fs ($10^{-15}$), the length of the light bullet is $\Delta l = c\Delta t = 3 \times 10^8 \Delta t$ or 0.3 m, 0.3 mm, and 0.3 $\mu$m, respectively. The bullet for the modern AK74 of caliber $5.45 \times 39$ mm has muzzle velocity 900 m/s and muzzle energy 1.39 kJ. For a light bullet of comparable length and energy, we require a pulse length of $\Delta t = \text{distance/speed} = 39 \times 10^{-3}/(3 \times 10^8) = 0.13$ ns or 130 ps with peak power $P_{\text{peak}} = \text{energy/time} = 139 \times 10^3/(0.13 \times 10^{-9}) = 10^{13}$ W, which is within the reach of a Q-switched Nd:YAG laser described in this chapter.

8. *Differences Between Light Bullet and Actual Bullet*: There are significant dif-ferences between a light and an actual bullet. The light bullet has a Gaussian profile, that is, higher intensity at the center. Depending on the wavelength, the light bullet may perform poorly in bad weather (Section 16.1), but in good weather, it may lose less power during propagation than an actual bullet and be so fast ($3 \times 10^8$ m/s) that movement of the target is immaterial. The laser has a faster firing repetition rate because it can operate 100 times per second. Light power can be adjusted for range to save power. In fact, to meet the Geneva convention of not blinding with a laser in battle, dazzle lasers must have their power controlled based on a range finder. The light bullet will interact with

bulletproof vests differently from a real bullet. Lasers are used to drill holes in ceramics, but reflective surfaces reduce effectiveness. Dragon plate armor, which spreads the bullet force over a wider area, is not likely to be effective against laser bullets. Explosive charges can be used to produce a light pulse in place of firing a bullet.

## 9.2 MODE-LOCKED LASERS

Lasers are generally longer than half the wavelength of the light they generate. Consequently, there are many longitudinal modes resulting from the Fabry–Perot resonances between the reflectors (Section 6.2 and Figure 6.13). The resonances in frequency are shown in Figure 6 14 The mode is determined by the number of half-cycles resonating, (Figure 6.12).

The frequency spacing $\Delta \omega$ between modes, (Figure 6.14), can be determined using equation (6.23) by writing angular frequency at two adjacent modes, with $n$ the mode number, $\eta$ the refractive index, and $d = n\lambda/2$ the length of the laser cavity: (using $f_n = c/\lambda_n$)

$$\omega_n = 2\pi f_n = \frac{2\pi c/\eta}{\lambda_n} = -\frac{2\pi cn}{2d\eta} = \frac{\pi cn}{\eta d}$$

$$\omega_{n-1} = \frac{\pi c(n-1)}{\eta d} \tag{9.1}$$

Subtracting the equations in equation (9.1),

$$\Delta \omega = \omega_n - \omega_{n-1} = \frac{\pi c}{\eta d} \tag{9.2}$$

Hence, the frequency separation between modes is for air $\eta = 1$

$$\Delta f = f_n - f_{n-1} = \frac{\Delta \omega}{2\pi} = \frac{c}{2d} = \frac{1}{2\tau} \tag{9.3}$$

where $\tau = d/c$ is the one-way travel time across the cavity (also called etalon).

### 9.2.1 Mode-Locking Lasers

Using exponential representation (the real part of which provides instantaneous time-dependent field), the electric field for $N$ modes can be represented by

$$E(t) = A \sum_{n=0}^{N-1} \exp\{-j(\omega_n t + \delta_n)\} \tag{9.4}$$

where for the $n^{\text{th}}$ mode, $w_n$ is angular frequency and $\delta_n$ is its phase, and we assumed a flat gain curve over the modes for simplicity.

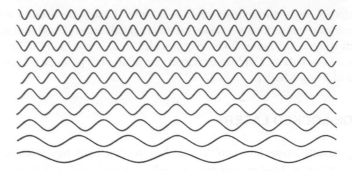

**FIGURE 9.1**   Diagram showing modes aligned at the center for mode locking.

As modes have different frequencies, power in equation (9.4) adds only incoherently. Then for each mode with intensity $A^2$, the intensity for $N$ modes is

$$I = NA^2 \tag{9.5}$$

Furthermore, the output is a continuous wave.

Mode locking involves adjusting the phase $\delta_n$ of each mode so that all modes have a peak at the same instant of time $t_0$ and its multiples at the center in Figure 9.1 for 10 modes. At multiples of time $t_0$, the modes will add coherently to provide a much larger peak than without mode-locking. The result is that the continuous wave is bunched into a series of pulses of peak power much greater than that of the continuous wave. Summing the 10 mode signals shown in Figure 9.1 produces pulses at multiples of $t_0$ in time (Figure 9.2). The mode signals in Figure 9.1 may be thought of as the Fourier decomposition of the pulses. Because of the cyclic nature of the resonances in time, the peaks of the waves will line up periodically, as seen in Figures 9.1 and 9.2. For $N = 10$ modes, the peak power of the mode-locked pulse is 10 ($N$) times the combined power for the $N = 10$ modes when they are not mode locked.

We now provide equations to prove this observation. Assuming that the modes are synchronized with $\delta = 0$ in equation (9.4) and that the incremental frequency between modes, $\Delta\omega$, is small relative to the nominal frequency $\omega_0$, letting

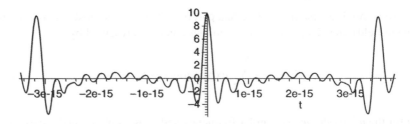

**FIGURE 9.2**   Mode locking sums modes in Figure 9.1 to produce a pulse string.

$\omega_n = \omega_0 + n\Delta\omega$, and neglecting $e^{-j\omega_0 t}$,

$$E(t) = A \sum_{n=0}^{n=N-1} e^{-jn\,\Delta\omega t} \qquad (9.6)$$

From equation (6.23), for the $n$th mode having frequency $f_n = \omega_n/(2\pi)$ and refractive index in the cavity $\eta$,

$$d = n\frac{\lambda_n}{2} = \frac{n}{2}\frac{c/\eta}{f_n} = \frac{n\pi c}{\eta\omega_n} \quad \text{or} \quad \omega_n = \frac{n\pi c}{\eta d} \qquad (9.7)$$

which is identical to the first equation in equation (9.1).

The spacing between mode frequencies (from mode $n$ to mode $n+1$) is from equation (9.2), $\Delta\omega = \pi c/(\eta d)$. For equation (9.6), there is a closed-form solution for $a < 1$:

$$\sum_{n=0}^{N-1} a^n = \frac{1 - a^N}{1 - a} \qquad (9.8)$$

Applying equation (9.8) to equation (9.6),

$$E(t) = A\left(\frac{1 - e^{-jN\,\Delta\omega t}}{1 - e^{j\,\Delta\omega t}}\right), \qquad (9.9)$$

Using $(1 - \exp\{j\,\Delta\omega t\}) = \exp\{j\Delta\omega t/2\}\,(\exp\{-j\Delta\omega t/2\} - \exp\{j\Delta\omega t/2\}) = \exp\{j\Delta\omega t/2\}(-j2\sin(\Delta\omega t/2))$, the power from equation (9.9) is

$$P_t = E(t)E^*(t) = A^2\frac{\sin^2(N\,\Delta\omega t/2)}{\sin^2(\Delta\omega t/2)} \qquad (9.10)$$

For small deviations in times away from the peak at $t = 0$, $\sin\theta \to \theta$, and the peak intensity can be approximated from equation (9.10):

$$A^2\frac{(N\Delta\omega t/2)^2}{(\Delta\omega t/2)^2} = A^2 N^2 \qquad (9.11)$$

Equation (9.11) shows a peak power $A^2 N^2$ that is $N$ times greater than the power $A^2 N$ for the nonmode-locked laser with $N$ modes from equation (9.5).

## 9.2.2 Methods of Implementing Mode Locking

Mode locking has the effect of bunching the output of a continuous wave laser into a string of pulses. When modes are locked, a pulse moves back and forward between

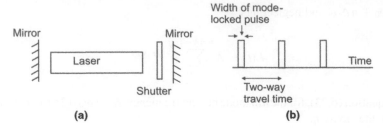

**FIGURE 9.3**   Implementing mode locking: (a) shutter in cavity and (b) shutter timing.

mirrors in the laser cavity. Mode locking can be accomplished by opening a shutter in the cavity long enough to allow the mode-locked pulse to pass through each time it crosses the cavity (Figure 9.3a). The pulses that open the shutter are periodic at a period of twice the travel time of the pulse across the cavity between the mirrors (Figure 9.3b). Only the locked pulse will resonate and build up energy in the cavity resonator, and any nonlocked arrangement of modes will be blocked by the shutter.

## 9.3   Q-SWITCHED LASERS

As an alternative to mode locking and not requiring many modes, a string of pulses of high peak power may be obtained from a laser by Q-switching. In Q-switching, propagation across the cavity is blocked while laser pumping continues to invert carriers to the upper level. Then, when the shutter releases the excess carriers in the upper level, they drop abruptly to generate an excessive number of photons in a very short time, creating a very powerful pulse. Figure 9.4 shows, as a function of time, the shutter opening, the fraction of excited atoms at the upper level, and the output pulse. While the shutter is closed, the laser pump continues to move atoms to the excited upper state, as seen in Figure 9.4b, way past the steady state that would have been reached without the shutter. As the shutter opens (Figure 9.4a), excess excited atoms rapidly drop to the ground state, generating many more photons than would have been generated in the steady state without the shutter. As a result, the excited

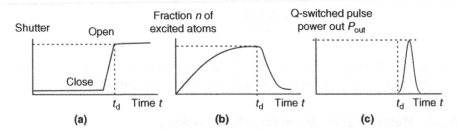

**FIGURE 9.4**   Timing for Q-switched laser: (a) shutter, (b) fraction of excited atoms at the upper level and (c) output pulse.

atoms in the upper level overshoot to a much lower number. Figure 9.4c shows the large output pulse generated.

The shutter can be implemented with an electro-optic switch, controlled by a voltage pulse. More commonly, a saturable absorber is placed in the cavity. A saturable absorber blocks light transmission until the light power exceeds a threshold. This avoids the need for timing. Note that saturable absorbers perform opposite to blockers in shields or goggles that block light that is too intense and could cause eye damage.

If $n_i$ and $n_t$ are the percentage of carriers per unit volume, initial and final, respectively, $N_0$ is the total number of energized atoms, $V$ is the volume of the cavity, $h$ is Plank's constant, and $v$ is the frequency associated with the bandgap of the material, then the total energy in a Q-switched laser pulse may be written as

$$E = \frac{1}{2}(n_i - n_t)N_0 V h v \qquad (9.12)$$

If $t_1 = 2d/c$ is the round-trip time and $R$ is the mirror reflectivity, then $1 - R$ of the light passes out of the cavity and the cavity lifetime for a photon may be written as

$$t_c = \frac{t_1}{1 - R} \qquad (9.13)$$

As a result, peak power can be computed from equations (9.12) and (9.13) as

$$P_m \approx \frac{E}{2t_c} \qquad (9.14)$$

## 9.4   SPACE AND TIME FOCUSING OF LASER LIGHT

Continuous wave light from a laser can be concentrated into a sequence of pulses in time by mode locking [176] (Section 9.2).

### 9.4.1   Space Focusing with Arrays and Beamforming

Coherent light from a group of lasers can be concentrated in space by synchronizing an array of lasers or splitting a source into many copies. In Figure 9.5, a linear 1D array of lasers is shown steered to a point $Q$ for illustrative purposes [162]. For $M$ lasers, if they are not synchronized, the average power at point $Q$ is $M$ times that of a single laser. The power at point $Q$ can be increased a further $M$ times by arranging the peaks from each laser to arrive at point $Q$ at the same time. This is accomplished by delaying or advancing the phases from the lasers so that the phase front points in the direction of $Q$ as shown. From Figure 9.5, to focus at $Q$, the extra distance that

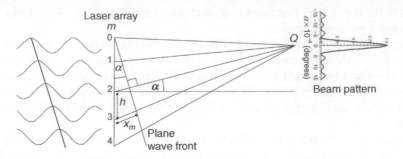

**FIGURE 9.5** Diagram showing how array processing produces a pulse at a specific angle in space.

the $m$th laser beam must travel relative to the zeroth laser is approximately

$$x_m = mh \sin \alpha \tag{9.15}$$

where $h$ is the distance between lasers in the linear array and $\alpha$ is the pointing angle. Therefore, the additional time advance for successive lasers is

$$\Delta t(\alpha) = \frac{x_{m+1}}{c/\eta_p} - \frac{x_m}{c/\eta_p} = \frac{h \sin(\alpha)}{c/\eta_p} \tag{9.16}$$

where, if a fiber delay line is used, $\eta_p$ is the refractive index of the material for the delay line. For narrow frequency band $\omega_0$, the delay corresponds to a phase of $\Delta t(\alpha)\omega_0$, which provides the correct delay in time for nominal wavelength $\omega_0$: the phase may be adjusted to steer the beam.

The field at $Q$, synchronized from all $M$ lasers, is then

$$E(\alpha) = A \sum_{m=0}^{m=M-1} e^{-jm \, \Delta t(\alpha)\omega_0} \tag{9.17}$$

By analogy with equation (9.6) and its closed-form equation (9.9) for mode locking in time, we can write the field for the spatial beam in closed form as

$$E(\alpha) = A \left( \frac{1 - e^{-jM \, \Delta t(\alpha)\omega_0}}{1 - e^{j\Delta t(\alpha)\omega_0}} \right) \tag{9.18}$$

By analogy with the intensity from equation (9.10) for mode locking in time, we can write the intensity for the beam pattern in space from equation (9.18) as

$$P_\alpha = E(\alpha)E^*(\alpha) = A^2 \frac{\sin^2(M \, \Delta t(\alpha)\,\omega_0/2)}{\sin^2(\Delta t(\alpha)\,\omega_0/2)} \tag{9.19}$$

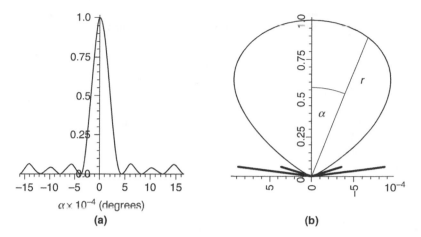

**FIGURE 9.6** Plots of intensity versus beam angle for a 1D array: (a) rectangular coordinates and (b) polar coordinates.

with $\Delta t(\alpha)$ given by equation (9.16). The power versus beam steering angle is shown in rectangular coordinates in Figure 9.6a and in polar coordinates in Figure 9.6b. The latter is referred to as the array beam pattern.

Note that in a similar manner as for time in equation (9.11), for small spacing $h$ between array elements, $\sin\theta \rightarrow \theta$. This approximates a peak power

$$P(\alpha)_{peak} = A^2 \frac{(M\Delta t(\alpha)\omega_0/2)^2}{(\Delta t(\alpha)\omega_0/2)^2} = A^2 M^2 \qquad (9.20)$$

which is $M$ times larger (due to focusing an array) than the nonsynchronized case for which the intensity is $A^2 M$. The analogy between time and space follows because both may be regarded as dimensions in a four-dimensional space $(x, y, z, ct)$. In practice, we use a 2D array with $M$ lasers for which a similar computation may be used.

## 9.4.2 Concentrating Light Simultaneously in Time and Space

In order for the light to arrive and add coherently at the same point in time and space at an angle $\alpha$, we synchronize both the mode-locked pulses from different lasers to arrive together at a point $Q$ in direction $\alpha$ and the light carrier frequencies between different lasers to be in phase at point $Q$.

### 9.4.2.1 Arranging Mode-Locked Pulses to Arrive Simultaneously at the Same Point
The mode-locked pulses can be controlled in time by adjusting the timing for the mode-locked shutters (Figure 9.3). Figure 9.7 shows a computer controlling the $m\Delta T$ time delay for the $m$th mode-locked laser.

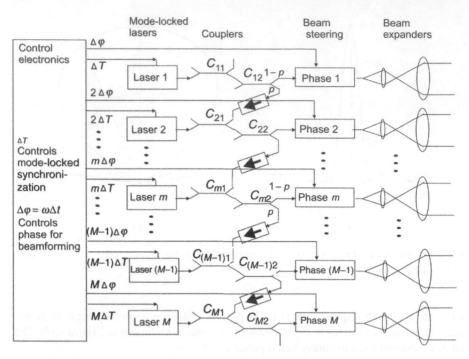

**FIGURE 9.7**   Array and mode-locked space and time focusing with steering.

#### 9.4.2.2   *Synchronizing the Light Carrier Frequency*   In Figure 9.7, fiber pigtailed laser diodes couple directly into the first column of fiber-connected optical directional couplers, $C_{11}$ to $C_{M1}$. For the couplers, the first subscript denotes the laser number from $m = 1, \ldots, M$ and the second subscript the coupler column number, 1 or 2, as shown in Figure 9.7 [57, 89, 106].

The synchronization of lasers at light frequencies in an array to perform beam-forming for focusing is achieved by splitting a small percent $p$ of the light emitted by the $m$th coupler in column 2 into the output of a successive $(m + 1)$th laser in column 1. An isolator is placed in this feedback path to ensure that the $(m + 1)$th laser is synchronized to follow the frequency and phase of the $m$th laser [110]. The percentage of light not coupled back $(1 - p)$ passes to a phase controller (or fiber delay line). The phases are set according to $\Delta t(\alpha)\omega_0$, where $\Delta t(\alpha)$ is determined by equation (9.16) to steer the beam in a fixed direction $\alpha$ from the normal to the array.

A beam expander (reverse telescope) (Section 1.3.4) provides a collimated beam that allows beams from all lasers to overlap at the array focus point. As the output beam from a single laser tends to be Gaussian shaped in space with a small radius of curvature, the optics is designed to focus the beam to its minimum waist (Section 2.1). For a broader frequency band, we should replace the phase controller with a time delay, such as a fiber delay, line with delays $\Delta t(\alpha)$.

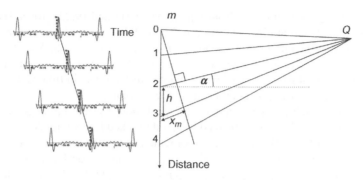

**FIGURE 9.8** Combining mode-locked lasers in an array.

### 9.4.2.3 Equations and Simulation for Concentrating Laser Light in Time and Space

Using the results of concentrating light in time and 1D space separately from Sections 9.2.1 and 9.4.1, respectively, we now combine these to synchronize $M$ mode-locked lasers, each having $N$ modes as shown in Figure 9.8. The field at a point $Q$ in space is obtained by combining the time and space field equations (9.6) and (9.17), respectively.

$$E(\alpha, t) = A \sum_{n=0}^{n=N-1} \sum_{m=0}^{m=M-1} e^{-j(m\,\Delta t(\alpha)\,\omega_0 - n\,\Delta\omega\,t)} \qquad (9.21)$$

The power $P(\alpha, t)$ at point $P$ for a given angle $\alpha$ and a specific time $t$ is computed from

$$P(\alpha, t) = E(\alpha, t)E^*(\alpha, t) \qquad (9.22)$$

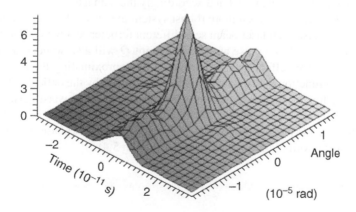

**FIGURE 9.9** Field concentrated in space and time.

Power at point $Q$ in Figure 9.8 from equations (9.21) and (9.22), for a 1D array, is plotted versus beam angle $\alpha$ and time in Figure 9.9 for a semiconductor laser of cavity length $d = 0.5$ cm, laser waveguide refractive index $\eta = 3.5$, $N = 10$ modes, $M = 5$ lasers in a linear array, a separation between lasers $h = 3$ cm, and a nominal wavelength $\lambda = 1.04$ μm. The half-power widths of the pulse in time and space are those for combining laser pulses at the array focus point $Q$.

For $M$ lasers, each having $N$ modes, the peak power at the target is $MN$ times greater than that for $M$ nonbeamformed nonmode-locked lasers with $N$ modes each. For $N = 50$ and a $10 \times 10$ array of lasers for which $M = 100$ lasers, the peak is 5000 times greater than that for the nonbeamformed array of nonmode-locked lasers. In the case of 2 W laser diodes in an array, this would provide $5000 \times 2$ W or 10 kW peak power. It is possible to increase the peak power further by using a longer laser, such as a water jacket cooled fiber laser, that exhibits more longitudinal modes. Because of beamforming from an array of lasers, the light intensity at less than or more than the range to point $Q$ is less for the laser array than for a single laser of identical power to the total array. This avoids collateral damage of a weapon system, reduces enemy ability to compromise a communication system, and provides better regional selection for spectrometry.

### 9.4.2.4 *Impact of Atmospheric Turbulence*

The pulse at $Q$ can be severely distorted in both space and time by atmospheric turbulence: the extent of distortion and subsequent peak power reduction depends on the strength of atmospheric turbulence [4, 155] (Section 5.4). For a single laser path, turbulence will spread the pulse in time [87] and generate multipaths for which the varying path lengths cause constructive and destructive interference with time—the reason a star twinkles. The beam expanders in Figure 9.7 expand the diameters of the laser beams, which also reduces the effects of turbulence by averaging over a larger cross section [4, 49, 155]. The optimum beam size and target area are application dependent, for example, in a laser weapon, a larger beam diameter will reduce the influence of turbulence but decrease the intensity at the target and accordingly the damage.

An array of lasers provides a more robust system than a single laser against multipath fading because each laser beam sees different turbulence. Owing to diffraction and turbulence [4, 48], the pulse at the focus point $Q$ will also wander around and expand in space. This effect may be computed by approximating the output beam with a Gaussian function of standard deviation determined by the optical output aperture and the radius of curvature determined to place the beam waist between laser and target. Turbulence is accounted for by computing phase masks as described in Refs [48, 88, 90, 144] and diffraction by propagating between phase masks using a Fourier transform formulation for diffraction (Section 5.4).

# CHAPTER 10

# ULTRAHIGH-POWER CYCLOTRON MASERS/LASERS

In Chapter 7, we described stimulated emission by bremsstrahlung for lasers having conventional bound state electrons. In this chapter, we consider higher power generation of stimulated emission by using cyclotron (or gyrodevice) concepts. We include cyclotron microwave masers with cyclotron optical lasers because of the overlap in technologies, for example, the similarity of the cyclotron lasers and masers, and because significant new applications fall between microwaves and optical frequencies. Note that the most promising of the cyclotron lasers and masers is the free-electron laser. Because of its importance, we dedicated Chapter 11 to the free-electron laser.

In Section 10.1, we discuss the relationship between conventional microwave devices and the much more powerful cyclotron devices that use bremsstrahlung to emit high-frequency electromagnetic radiation. In Section 10.2, we describe the principles and operation of the basic continuous wave gyrotron, normally operating at high-frequency microwaves. In Section 10.3, we describe the operation and a tabletop laboratory demonstration of a vircator that uses the cyclotron effect for robust simple generation of high-power pulses on a high-frequency carrier, usually microwaves, for pulsed applications (Chapter 9).

*Military Laser Technology for Defense: Technology for Revolutionizing 21st Century Warfare*,
First Edition. By Alastair D. McAulay.
© 2011 John Wiley & Sons, Inc. Published 2011 by John Wiley & Sons, Inc.

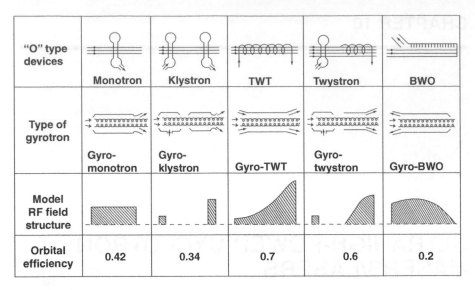

| | | | | | |
|---|---|---|---|---|---|
| "O" type devices | Monotron | Klystron | TWT | Twystron | BWO |
| Type of gyrotron | Gyro-monotron | Gyro-klystron | Gyro-TWT | Gyro-twystron | Gyro-BWO |
| Model RF field structure | | | | | |
| Orbital efficiency | 0.42 | 0.34 | 0.7 | 0.6 | 0.2 |

**FIGURE 10.1** Comparison gyrodevice with conventional counterparts.

## 10.1 INTRODUCTION TO CYCLOTRON OR GYRO LASERS AND MASERS

Microwave tubes have been around for many years and were crucial for the significant impact of radar in World War II. They are widely used in microwave ovens and industrial heating today and are a core part of microwave curriculum in electrical engineering departments. In these tubes, the electron cyclotron interaction between electrons and electromagnetic field is limited to electromagnetic fields with slow phase velocity by using cavity entrances (kylstron and some magnetrons) and surface waves from periodic structures, traveling-wave tubes (TWT), and backward wave oscillators (BWO) [24]. The generation of slow waves is local at periodic structures or cavity entrances, which limits the interaction space and hence the power that can be generated. Such devices are known as ordinary or O-type devices (Figure 10.1) [65].

We focus on devices that use fast phase velocity waves whose relatively unrestricted interaction space allows orders of magnitude higher power. For a fast velocity version of a slow wave device, we add the prefix gyro- (Figure 10.1). A gyroklystron millimeter wave amplifier at 94 GHz, an atmospheric window in W-band between optics and microwaves, is described in Chapter 16 for radar and high-resolution mapping through increment weather, such as for mapping of space debris in lower earth orbit. Furthermore for airborne applications, we may choose to use pulsed sources because the average power can be low (100 W), while the peak power can be gigawatts.

### 10.1.1 Stimulated Emission in an Electron Cyclotron

In the electron cyclotron laser or maser, a static electromagnetic field causes a stream of electrons in a vacuum tube to bunch up so as to create higher and lower electron densities that act as repetitive high and low electron energy levels. Bunching occurs because, from Einstein's theory of special relativity, relativistic electrons (those with sufficient speed for relativity to matter) increase mass when traveling faster. This slows them down so that now their mass decreases and they speed up again. This sets up an oscillatory pattern or bunching of electron density in space. Electrons can be stimulated to fall from a high to low energy levels (of which there may be many) to generate radiation at a frequency related to the bunching. As in the bound electron state case, the coherent electromagnetic radiation produced will be identical to that provided by a seed, but, unlike the bound electron state case, the frequency generated is not limited to the material bandgap.

The input seed field helps initiate oscillation in a structure with a resonant cavity by focusing energy in a narrowband. The quality factor $Q$ is $2\pi$ times the energy stored divided by energy lost per cycle, where energy stored per unit volume is proportional to $0.5n^2E^2$, where $E$ is the electric field and $n$ is the refractive index of the medium. High reflectivity at the ends of the cavity results in a sharp peak in frequency corresponding to a narrowband with high power and high $Q$ (Section 6.2). However, this makes switching slower as there is more energy in the cavity that has to change to a new frequency: switching time is proportional to $Q/\omega$. This makes the creation of short pulses more difficult.

In radar applications such as that considered in chapter 16, the maser or laser is often used as an amplifier because the optimized radar transmit signal is complicated and can more easily be generated in electronics and then amplified. The seed becomes the input field for an amplifier. Design parameters vary according to whether an amplifier or an oscillator is desired. Two varieties of electron cyclotron maser are discussed.

1. Gyrotrons, or cyclotron resonance masers (CRM), (Section 10.2), have a stream of relativistic electrons in a smooth-walled evacuated tube with a static external magnetic field along the tube.
2. If an electrostatic field replaces the magnetic one, the electrons oscillate in a potential well and the pulsed device is called a vircator or reditron (Section 10.3).

In Chapter 11 we consider a gyrotron with a periodic external field called a free-electron laser or a free-electron maser. Free-electron lasers and masers are discussed in a separate chapter because of their importance at light frequencies. Gyrotrons tend to be used continuously or with long pulses. So they must be followed by a pulse forming network if a pulse is required. (In contrast, the vircator is an impulse source.)

## 10.2 GYROTRON-TYPE LASERS AND MASERS

Gyrotrons and gyroklystrons are widely used for efficient continuous or long-pulse generation of high-power microwaves (peak power greater than 100 MW or continuous power over 1 MW). For example, in Chapter 17, gyrotrons and gyroklystrons are used for active denial, body scanners, inspecting packages, destroying electronics, and in Chapter 16 for high-resolution radar in inclement weather.

### 10.2.1    Principles of Electron Cyclotron Oscillators and Amplifiers

In electron cyclotron maser stimulated emission [129], the interaction is between relativistic electrons (close to the speed of light $c$) and an electromagnetic field with fast phase velocity (exceeding the speed of light). The stream of electrons is at a slight angle to the axis of the magnetic field. We now show that this causes the electrons to gyrate in a helical path around the axis. First, we neglect the relativistic nature of electrons. For motion of a charged particle in a constant and uniform magnetostatic field, from Newton's law, mass times acceleration equals force [13]:

$$m\frac{d\mathbf{v}}{dt} = q(\mathbf{v} \times \mathbf{B}) \tag{10.1}$$

where $\mathbf{B}$ is the magnetic field flux density, and a particle has rest mass $m$, velocity $\mathbf{v}$, and charge $q$. Splitting $\mathbf{v}$ into components $\mathbf{v}_{\parallel}$ that is parallel to $\mathbf{B}$ and $\mathbf{v}_{\perp}$ that is perpendicular to $\mathbf{B}$,

$$\mathbf{v} = \mathbf{v}_{\parallel} + \mathbf{v}_{\perp} \tag{10.2}$$

Figure 10.2b shows the case with axes rotated so that $\mathbf{v}$ is in the $\hat{x}$–$\hat{z}$ plane of Figure 10.2a. In the cross product in equation (10.1), only $\mathbf{v}$ components at right angles to $\mathbf{B}$ are nonzero. So equation (10.1) becomes

$$\frac{d\mathbf{v}_{\perp}}{dt} = \frac{q}{m}(\mathbf{v}_{\perp} \times \mathbf{B}) \tag{10.3}$$

If we define a vector

$$\mathbf{\Omega}_{c} = \frac{-q\mathbf{B}}{m} \tag{10.4}$$

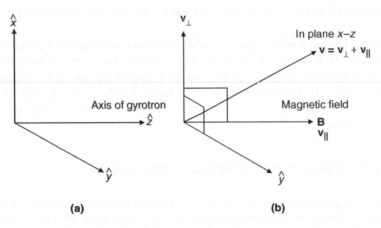

**FIGURE 10.2**    Splitting velocity $\mathbf{v}$ into parallel and perpendicular components: (a) coordinate system and (b) components of velocity.

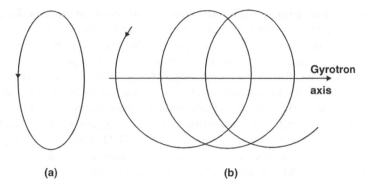

(a) (b)

**FIGURE 10.3** Resulting trajectory of electron in gyrodevice: (a) transverse circular trajectory and (b) resulting helical trajectory combines $v_\parallel$ and transverse circular.

then equation (10.3) becomes

$$\frac{d\mathbf{v}_\perp}{dt} = \boldsymbol{\Omega}_c \times \mathbf{v}_\perp \tag{10.5}$$

From equation (10.4), $\boldsymbol{\Omega}_c$ is in the direction of **B**. Therefore, equation (10.5) suggests that the electron accelerates in a direction at right angles to the plane containing **B** and $\mathbf{v}_\perp$. Acceleration in the $y$ direction in Figure 10.2b causes $\mathbf{v}_\perp$ to rotate around the $z$ axis, and hence the electron to follow a circular path in the transverse plane (Figure 10.3a). As the particle also has velocity $v_\parallel$ along the guide, the result is a helical trajectory around the tube axis as shown in Figure 10.3b. The magnitude $\Omega_c = |\boldsymbol{\Omega}_c|$ from equation (10.4) can be written as

$$|\boldsymbol{\Omega}_c| = \frac{|q||\mathbf{B}|}{m} \quad \text{or} \quad \Omega_c = \frac{|q|B}{m} \tag{10.6}$$

and $\Omega_c$ is known as the angular frequency of gyration, gyrotron frequency, cyclotron frequency, or Larmor frequency.

For relativistic electrons, the mass depends on velocity, so replacing $m$ by $mc\gamma$, equation (10.4) becomes

$$\Omega_c = \frac{|q|B}{mc\gamma} \tag{10.7}$$

where the relativistic factor

$$\gamma = \frac{1}{\sqrt{1 - v^2/c^2}} \tag{10.8}$$

An important conclusion from equations (10.7) and (10.8) is that the gyro or cyclotron frequency $\Omega_c$ depends on magnetic field $B$ and on the energy of the electrons from the accelerator (electron gun) through $\gamma$ and hence $v$ and also influences the frequency of the electromagnetic oscillation (laser light) produced by electrons moving at relativistic velocity $v$ in a constant uniform magnetic field. Therefore, the frequency of a gyrodevice can be tuned by changing the voltage on the electron gun, hence changing the electron velocity $\mathbf{v}$; this allows a large range of frequencies $\Omega_c$. This contrasts the bound electron state case for conventional lasers (such as laser diodes) and masers for which the frequency is determined by the material bandgap that is extremely limiting.

A second effect, bunching of relativistic electrons, produces repetitive high–low energy levels that support stimulated emission. For a TEM mode propagating down the waveguide, the electric field $(E)$ is transverse. Therefore, at one side of the circular electron path, the $E$ field contributes a force $qE$ to accelerate the electrons, while on the other side of the circular path, the $E$ field decelerates the electron. The combination of acceleration and deceleration results in bunching of electrons to produce higher density and lower density periodicity. This periodicity of electron bunching enables stimulated emission. Very large power is possible because of absence of a nonlinear material and opportunity for large interaction volume when using a large diameter smooth vacuum tube. The result is orders of magnitude higher power than that for nonrelativistic electrons or slow phase microwave tubes. In addition, the high power is produced very efficiently and can be used to generate coherent microwaves over a large range of frequencies. The upper end of the range approaches optical frequencies.

## 10.2.2  Gyrotron Operating Point and Structure

The gyrating electrons not only acquire an orbital velocity but also change their speed $v_z$ in the axial direction to produce a Doppler effect [129]. Therefore, the gyration frequency $\Omega$ differs from the angular frequency $\omega$ of the electromagnetic wave according to

$$\omega - k_z v(z) \approx s\Omega \tag{10.9}$$

where $s$ allows harmonics of $\Omega$.

The gyrotron operates by interaction between the gyrating electrons and the electromagnetic wave. Figure 10.4 [9] shows a dispersion diagram, angular frequency $\omega$ versus propagation constant $k_z$. For the cyclotron wave from gyrating electrons, the graph of equation (10.9) is a straight line as shown in the figure. The microwave field in the waveguide is represented by a parabola for the normal TE waveguide mode:

$$\omega^2 = \omega_{co}^2 + c^2 k_z \tag{10.10}$$

where $\omega_{co}$ is the cutoff angular frequency. The intersection of the two curves, equations (10.9) and (10.10), provides the operating point for the gyrotron resulting from the interaction of the gyrating electrons and the microwave field and gives the frequency of oscillation that will be observed. Two different variations of gyrotron are shown with different operating points. A gyrotron traveling-wave tube (gyro-TWT) provides forward velocity of the electromagnetic wave. Traveling wave means that

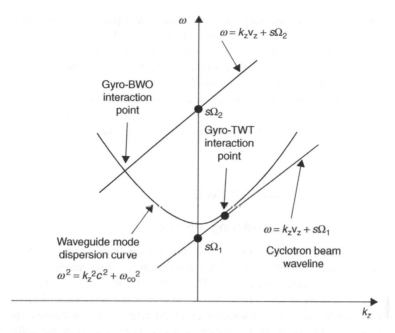

**FIGURE 10.4** Dispersion diagram showing operating point for gyrotron.

the electromagnetic field is not reflected back and forth in a cavity, so it is used for amplification with frequency $\Omega_1$ rather than as an oscillator. The parameter $s = 1$ for the fundamental frequency $\Omega_c$. A gyro-backward wave oscillator (gyro-BWO) has the electromagnetic field moving in the direction opposite ($k_z$ negative) to the electron stream and gives an oscillating frequency of $\Omega_2$ exiting in reverse (Figure 10.1).

Figure 10.5 shows a typical gyrotron structure with efficiency greater than 40% [9, 129]. A voltage at the anode creates a field at the cathode. The magnetron-type electron gun is directed at a slight angle to the magnetic field $B$ axis created by the

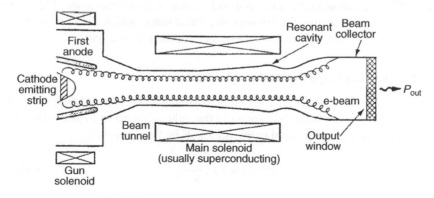

**FIGURE 10.5** Typical gyrotron basic structure.

main solenoid to produce electron velocity components parallel and perpendicular to the magnetic axis. Thus, electrons have axial and orbital components leading to a helical path around the axis (Section 10.2.1). Electrons move into growing $B$ field and are adiabatically compressed, causing orbital momentum to increase. In the uniform $B$ field, electrons interact with the eigenmodes of the electromagnetic microwaves in the cavity and some of the kinetic energy transforms to microwave energy. The spent beam exits from the axially open cavity, is decompressed in the decreasing magnetic field, and settles on the collector. The generated microwave passes through the output window.

## 10.3   VIRCATOR IMPULSE SOURCE

### 10.3.1   Rationale for Considering the Vircator

Compact, lightweight, efficient, high-power microwave sources for mobile military platforms have been the subject of extensive research for many years under a series of multi university research initiatives (MURIs) [9, 41]. While we could use almost any maser by following it with a high-power pulse forming network, this approach is generally not conducive to a compact, efficient, lightweight source because impedance matching components are needed between each step. For this reason, we believe that a direct impulse source will be more effective.

The virtual cathode oscillator (vircator) has been extensively studied over the years [3, 9, 40, 123] and is simple, robust, light, and among the most promising direct pulse generating compact sources, giving gigawatts of peak power. Its only weakness is that in the past it was less than 10% efficient. Current research is aimed at higher efficiency and frequency locking. The vircator can be lighter than gyrotrons and free-electron masers because it can use an electrostatic field instead of a magnetic field. Magnets are heavy and solenoids require another electrical source. The 1988 patent [70] references earlier patents and uses a magnet. Subsequent vircators and an improved version, the reditron, do not [27]. Furthermore, a tabletop electrostatic vircator demonstration unit was constructed and tested at below 5 GHz, suitable for mine and IED detection, providing real-world data on vircator performance [123]. The vircator is also robust against the driving power pulse shape, which allows inexpensive drivers such as the Marx generator without any impedance matching.

### 10.3.2   Structure and Operation of Vircator

At high voltage, several hundred kilovolts, between anode and cathode, electrons are emitted from the cathode in a stream to the right as shown in Figure 10.6a. The electron beam current $I$ is related to the voltage $V$ by the perveance

$$\text{Perveance} = \frac{I}{V^{3/2}} \tag{10.11}$$

A high perveance is more efficient [9].

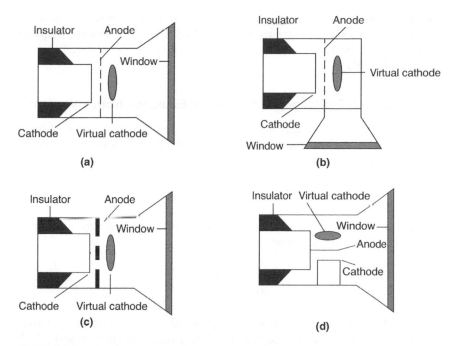

**FIGURE 10.6** Virtual cathode oscillator: (a) basic vircator, (b) transverse extracted, (c) reditron, and (d) reflex triode removes chirp.

The vircator operates by forming a virtual cathode, hence name virtual cathode oscillator (vircator). The virtual cathode arises when the cathode generated current exceeds the maximum current that can pass through the drift space of the tube. This current is called the space charge current and is of the order of 10 kA. If the injected beam current is greater than the space charge current and the radius of the drift space region is sufficient, the excess space charge forms a strong potential barrier at the head of the beam. This is the virtual cathode that reflects electrons to form an oscillating cavity. The cavity reinforces microwave radiation by bremsstrahlung radiation. The cavity can be designed to radiate at a frequency suitable for airborne mine detection (say less than 10 GHz). In the structures shown in Figure 10.6a and b [9], the virtual cathode oscillates axially, but in Figure 10.6b the direction of the microwave output beam is transverse to the axis.

When the drive voltage is switched off at the end of a pulse, the amplitude of the virtual cathode oscillation decays and the frequency increases as the cavity length reduces. This chirp significantly increases bandwidth that may be useful in some applications.

Some variations of the vircator are aimed at removing the chirp to maintain a constant frequency (narrowband), in order to maximize peak power. Figure 10.6c shows a reditron [27] that uses a thick anode plate with holes to remove virtual cathode oscillation and hence the chirp. The peak power of an experimental unit was 1.6 GW. Figure 10.6d shows a reflex triode that also eliminates the oscillation of the virtual cathode and hence the chirp. This is the configuration that was selected for the tabletop demonstration unit described in Section 10.3.5. By switching the frequency through a sequence of frequency steps such as 8.5, 9.5, 10.5, and 11.5 GHz, we can design a matched filter.

### 10.3.3 Selecting Frequency of Microwave Emission from a Vircator

In short pulse operation, the maximum level of radiated power is restricted by the microwave breakdown near the structure walls [40]. The breakdown restricts the maximum electric field according to

$$E_{br} = 0.8 \times 10^3 \sqrt{f} \tag{10.12}$$

From Newton's equation, the maximum frequency of oscillation near a positively charged electrode is provided from

$$\frac{d^2 z}{dt^2} = \frac{eE_z}{m\gamma^3} \tag{10.13}$$

where $eE_z$ is the force on a particle with charge $e$ and rest mass $m$ near a positively charged electrode, $E_z$ is the electric field along the axis of the vircator, and $\gamma$ is the relativistic factor. For a time harmonic, single angular frequency $\omega_0$, $(d^2 z)/(dt^2) = \omega_0^2 L$ with amplitude of oscillation $L = V/E_z$. Hence,

$$\omega_0^2 = \frac{eE_z}{m\gamma^3 L} = \frac{eE_z^2}{m\gamma^3 V} \tag{10.14}$$

At breakdown, $E_z = E_{br}$, where $E_{br}$ is the dielectric breakdown strength. Substituting equation (10.12) into equation (10.14) gives

$$\omega_0^2 = (2\pi f)^2 = \frac{e(0.8 \times 10^3)^2 f}{m\gamma^3 V} \tag{10.15}$$

or

$$f = \frac{1}{(2\pi)^2} \frac{e(0.8 \times 10^3)^2}{m\gamma^3 V} \approx 3 \times 10^{15} V^{-1} \tag{10.16}$$

Therefore, at 300 kV, the frequency is 10 GHz.

### 10.3.4 Marx Generator

The vircator needs a power supply of 300 kV and current greater than 10 kA with a rise time less than 100 ns. A cost-effective power supply can be constructed with a Marx generator. The Marx generator was patented in 1924 and is used to simulate

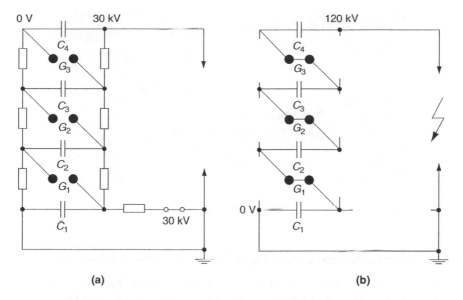

**FIGURE 10.7**    Marx generator structure: (a) showing charging and (b) discharge.

lightning, generate X-rays, and ignite a thermonuclear device. It charges a number of capacitors in parallel and then discharges them in series. We would like the output impedance of the Marx generator to match the input impedance of the vircator for maximum transfer of energy to the vircator load. This avoids large and expensive impedance matching components.

Figure 10.7a shows the structure of a four-stage Marx generator in which a 30 kV DC source charges four capacitors in parallel. For a system with $N$ stages, the $V_{charge} = 30$ kV power supply source charges all $N$ capacitors of value $C_{stage}$, $C_1$ to $C_N$, in parallel up to 30 kV through boxes containing resistors. The $N$ capacitors in parallel give an equivalent capacitance for charging of $C_{set} = NC_{stage}$ (Figure 10.7a). Inductors may be used in place of charging resistors for more efficiency and speed.

To generate a high-voltage pulse at the right-hand side of Figure 10.7b, a spark gap $G_1$ is triggered to form a short circuit between the two dark circles at $G_1$. This now doubles the voltage across $G_2$, causing it to trigger into a short circuit. This repeats rapidly until all the capacitors are discharged in series as shown in Figure 10.7b, where the peak voltage is $4 \times 30$ kV $= 120$ kV. In general, on discharge, a pulse is generated at the right-hand side of Figure 10.7b of peak approximately $V_{peak} = NV_{charge}$. For discharging, the $N$ capacitors in series have a capacitance of $C_{set} = C_{stage}/N$ and inductance $L_{set} = NL_{stage}$. Therefore, the Marx output impedance may be written as

$$Z_{Marx} = \sqrt{\frac{L_{set}}{C_{set}}} = N\sqrt{\frac{L_{stage}}{C_{stage}}} \qquad (10.17)$$

**FIGURE 10.8**   Photograph of combined Marx generator with vircator demonstration.

### 10.3.5   Demonstration Unit of Marx Generator Driving a Vircator

A vircator driven by a Marx generator appears to be capable of producing 200 MW impulses suitable for airborne applications such as ground penetration radar [123]. Figure 10.8 [123] shows a photograph of a tabletop demonstration unit in which a 25-stage Marx generator drives a reflex triode vircator to provide a high-power microwave impulse source suitable for airborne mine detection. The 18 Ω output impedance of the Marx generator is similar to the input impedance of the vircator, so large and lossy impedance matching components such as transmission lines, transformers, or tapered waveguides are avoided. This simplifies the system, reducing cost and weight and improving reliability. The maximum voltage that can be delivered to the matched load is $(1/2)N \times V_{\text{charge}}$. So for a charging voltage of $V_{\text{charge}} = 20$ kV and 25 stages, the maximum voltage is 250 kV.

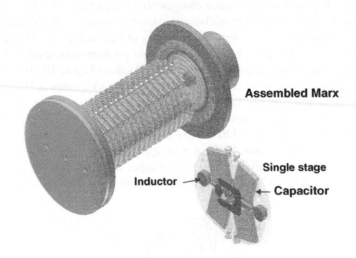

**FIGURE 10.9**   Photograph of Marx generator used in demonstration.

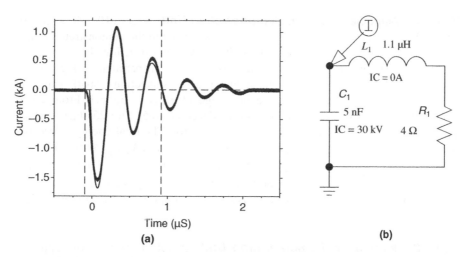

**(a)**

**(b)**

**FIGURE 10.10** 20-stage Marx generator test: (a) output pulse and (b) equivalent circuit.

A second advantage for the vircator is that it is forgiving with respect to driver voltage fluctuations compared to other masers and gyrodevices; most of these require a flattop pulse. Furthermore, no other power supply is needed because the vircator uses the electrostatic field and does not require a magnetic field. This further simplifies the system and reduces weight and cost and increases reliability.

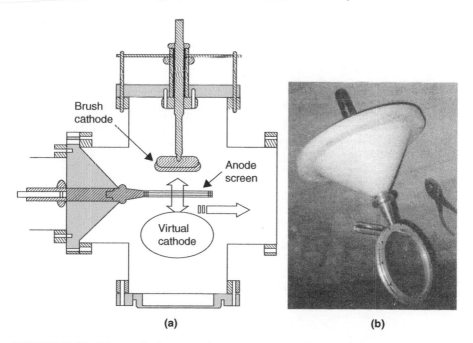

**(a)**

**(b)**

**FIGURE 10.11** Vircator in demonstration unit: (a) reflex triode-type vircator and (b) anode stalk.

**10.3.5.1   Marx Generator in Demonstration Unit**   Figure 10.9 shows a photograph [123] of the $N = 25$-stage Marx generator used in the demonstration without its steel cover: the cover is part of the coaxial construction. Each stage of the Marx generator has four mica capacitors of 25 nF in parallel, giving a stage capacitance of $C_{stage} = 100$ nF. The inductance of connections, spark gap, and capacitors add to give an inductance per stage of $L_{stage} = 55$ nH. From equation (10.17), these values give an output impedance of 18.5 $\Omega$ for 25 stages, which approximately matches the impedance of the vircator. The total energy stored before discharge is $E = (1/2)C_{set}V_{peak}^2$ where $V_{peak} = N \times V_{charge} \approx 500$ kV and $C_{set}(\text{series}) = C_{stage}/N \approx 4$ nF, giving $E = 0.5 \times 4 \times 10^{-9} \times (5 \times 10^5)^2 \approx 500$ J.

In an earlier test of the Marx generator with $N = 20$ stages and a low charging voltage of $V_{charge} = 1.5$ kV, the pulse out is shown in Figure 10.10a and the equivalent LCR circuit in Figure 10.10b [123].

**10.3.5.2   Vircator in Demonstration Unit**   A velvet cathode with approximately 6 cm diameter and a 10 mm gap between anode and cathode is shown in Figure 10.11a [71] for the reflex triode vircator [123]. The anode stalk is shown in Figure 10.11b [123]. This vircator, charged with 20 kV, gave 50 MW with a 220 kV peak voltage. The input impedance fluctuates starting around 18 $\Omega$ and dropping toward the end of the pulse; 5 GHz was generated. Researchers expect to get 200 MW with this design. The charging time of approximately 100 ms limits the repetition rate. Cooling will allow faster charging.

# CHAPTER 11

# FREE-ELECTRON LASER/MASER

We describe the free-electron laser in this chapter because it is the most promising of the ultrahigh-power cyclotron lasers. Chapter 10 provides the background for this chapter.

We expect the free-electron laser to revolutionize high-power generation of coherent light radiation for lasers that will in turn revolutionize the use of lasers in military applications including manufacturing, which is critical for military buildup in a crisis. The development of free-electron lasers is progressing steadily. In 1994, the Navy funded Jefferson Laboratory demonstrated 10 MW at 6 μm and in 2005 plans were underway for 10 MW at 1 and 3 μm, 1 MW at UV, and 100 W in terahertz range [170]. Since then, several large contracts have been awarded to construct large free-electron lasers. For example, a press announcement in June 2009 [126] indicated that Raytheon was awarded a 12-month contract with ONR to perform the preliminary design of a 100 kW experimental free-electron laser. The free-electron laser is considered for use with future millimeter radar (radio frequency detection and ranging) in Chapter 16 and lidar (light detection and ranging) in Chapter 15.

In Section 11.1, we describe the significance and principles of the free-electron laser/maser (FEL/FEM) that is distinguished from other cyclotron lasers by the addition of a wiggler whose spatially oscillating field enables the range of frequencies of the gyrotron to be extended through light up to X-rays. With variation it can also extend the range down to 0.28 GHz [33]. The frequency of the generated microwaves now depends on both the periodicity of this structure and the oscillating energy of the electrons. In Section 11.2, we explain the theory of free-electron operation to

*Military Laser Technology for Defense: Technology for Revolutionizing 21st Century Warfare,*
First Edition. By Alastair D. McAulay.
© 2011 John Wiley & Sons, Inc. Published 2011 by John Wiley & Sons, Inc.

show that electron bunching produces stimulated laser emission by bremsstrahlung and why any wavelength can be generated. In Section 11.3, we describe a proposed airborne free-electron laser [170] at high multimegawatt power and an inexpensive laboratory tabletop free-electron X-band laser [173, 175], with low multikilowatt power.

## 11.1 SIGNIFICANCE AND PRINCIPLES OF FREE-ELECTRON LASER/MASER

### 11.1.1 Significance of Free-Electron Laser/Maser

The reasons why free-electron lasers show the most promise for high-power lasers and masers are as follows [170]:

1. The wavelength of free-electron laser light can be chosen for a specific application from an impressive range from radio frequencies through IR and UV to X-rays. This contrasts bound electron state lasers covered in Chapters 7 to 9 for which the wavelength depends on the specific bandgap of the material. It contrasts gyrotrons that do not extend up through light frequencies.
2. Because of the free-electron laser flexibility in wavelength and tunability, the same design can match and cover diverse applications at significant cost savings.
3. With a free-electron laser, megawatts of light can be available at any time and within seconds in contrast to other high-power laser systems that generally have significant warm-up and/or recovery time when fired.
4. The free-electron laser light can be modulated so that matched filtering may be used with return signals in radar to enhance signal to noise ratio.
5. The free-electron laser is efficient—approximately 30% of the electronic power supplied is converted into light at megawatt power.
6. The free-electron laser is powered by electricity that can be generated from an aircraft engine and does not require storage of toxic chemicals or batteries. A multimegawatt laser on a 747 is expected to need less than 7% of the engine capacity.

### 11.1.2 Principles of Free-Electron Laser/Maser

As described in Chapter 10, the FEL/FEM generates or amplifies electromagnetic radiation in light or microwaves by converting energy in a stream of electrons into high-power (many megawatts) coherent continuous or pulsed electromagnetic energy at any frequency in microwaves (1–300 GHz) or light (300 GHz–UV).

In addition to the elements used in all high-power gyrodevices, electron accelerator (gun) generating relativistic electrons (close to the speed of light), fast phase velocity electromagnetic field, and absence of nonlinearity to limit power, the FEL has a

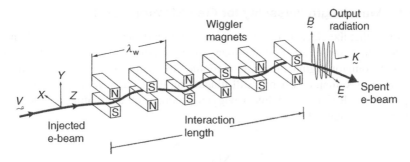

**FIGURE 11.1**    Principle for free-electron laser.

periodic magnetic field along the vacuum tube (Figure 11.1) [65]. The periodic magnetic field is provided by an external wiggler (also called inverter or undulator in other applications) and enables the FEL to distinguish itself from all other high-power gyrodevices. By external we mean that the components creating the periodic field in the vacuum tube are located outside the tube. Higher frequencies are achieved than in a gyrotron because the wiggler Doppler upshifts the frequency of the wiggler by approximately the relativistic parameter $\gamma^2$, equation (10.8). Normally, the advantage of a wiggler is the ability to generate higher frequencies. But the wiggler also provides additional design flexibility that allows rapid tuning over greater than 15%, extended frequency range up and down, and bandwidth control that can be critical for some applications regardless of frequency [65, 157]. Free-electron lasers have been designed to operate at frequencies as low as 280 MHz (1 m wavelength) [33].

While the same principles are used for FELs throughout the impressive range of frequencies, the implementation and analysis of the FEL may differ radically between lower and higher frequencies, for example, in electron accelerator (gun) technology and in cavity design. The energy in the electron beam is one of the factors used to control frequency. For high frequencies (FEL, $\lambda < 0.5$ mm), high electron energy ($> 10$ MeV) uses the Compton effect, while for low frequencies (FEM, $\lambda > 0.5$ mm), low beam energy ($< 10$ MeV) and higher current use the Raman effect. In Figure 11.1, an electron beam of relativistic electrons (velocity $v$ close to the speed of light $c$) from an electron gun is injected into a smooth-walled vacuum tube. A static periodic magnetic field along the tube acts as a wiggler or an inverter. The energy from the electrons is converted into electromagnetic energy in the form of microwaves or light. The spent beam exits.

## 11.2    EXPLANATION OF FREE-ELECTRON LASER OPERATION

We now show, for a free-electron laser, that electron bunching for stimulated bremsstrahlung emission occurs and that any wavelength light is obtained by adjusting the wiggler wavelength and the magnetic field strength.

## 11.2.1   Wavelength Versatility for Free-Electron Laser

We assume that the electron density is small enough during the exchange of energy from electron beam to radiation that many-electron collective effects are negligible. This is the Compton regime of operation for the FEL. This is easier to explain than the many-electron Raman case while still illustrating the general principles. We follow the approach in Ref. [119]. A linear wiggler produces an alternating magnetic field along the tube (Figure 11.1):

$$\mathbf{B}_\mathrm{w} = \hat{\mathbf{x}} B_\mathrm{w} \cos \frac{2\pi z}{\lambda_\mathrm{w}} \tag{11.1}$$

For a linear wiggler, the output electromagnetic wave is linearly polarized in the $\hat{\mathbf{x}}$ direction. The spatial frequency of wiggler field is $f_\mathrm{w} = 1/\lambda_\mathrm{w}$. Or we can use a helical wiggler

$$\mathbf{B}_\mathrm{w} = B_\mathrm{w} \left( \hat{\mathbf{x}} \cos \frac{2\pi z}{\lambda_\mathrm{w}} + \hat{\mathbf{y}} \sin \frac{2\pi z}{\lambda_\mathrm{w}} \right) \tag{11.2}$$

For a helical wiggler, the output field is circularly polarized. On electrons at velocity $\mathbf{v}$, the magnetic wiggler exerts a force

$$\mathbf{F}_\mathrm{w} = e\mathbf{v} \times \mathbf{B}_\mathrm{w} \tag{11.3}$$

where $e$ is charge of an electron. From equation (11.3), the force is transverse to the plane containing velocity $\mathbf{v}$ and magnetic field $\mathbf{B}_\mathrm{w}$ and there is no force on the electron traveling in direction $\mathbf{v}$, or $\mathbf{F}_\mathrm{w}\mathbf{v} = 0$. But the magnetic field exerts a transverse (sideways) force on the electron, causing it to wiggle back and forth as it passes through the periodic magnetic field (Figure 11.1).

The electromagnetic wave generated (or provided at input for an amplifier) has a transverse electric field $\mathbf{E}$, which can interact with the transverse oscillating electrons through a force $e\mathbf{E}$. The transfer of energy is used to amplify the electromagnetic wave.

The electron velocity is written as

$$\mathbf{v} = v_x \hat{\mathbf{x}} + v_y \hat{\mathbf{y}} + v_z \hat{\mathbf{z}} \tag{11.4}$$

where $v_z$ along the tube is close to the speed of light $c$ for relativistic electrons and the distance along the tube is $z = v_z t \approx ct$. Then, from Newton's equation for an electron, static mass $m$ times acceleration $\dot{v}$ equals force using equation (11.2) and (11.3)

$$m\dot{\mathbf{v}} = ec\mathbf{B}_\mathrm{w} \quad \text{or} \quad m(\dot{v}_x \hat{\mathbf{x}} + \dot{v}_y \hat{\mathbf{y}}) = ecB_\mathrm{w} \left( -\hat{\mathbf{x}} \sin \frac{2\pi ct}{\lambda_\mathrm{w}} + \hat{\mathbf{y}} \cos \frac{2\pi ct}{\lambda_\mathrm{w}} \right) \tag{11.5}$$

So

$$\dot{v}_x = -\frac{e}{m} B_w c \, \sin \frac{2\pi ct}{\lambda_w} \quad \text{and} \quad \dot{v}_y = \frac{e}{m} B_w c \, \cos \frac{2\pi ct}{\lambda_w} \tag{11.6}$$

Integrating equation (11.6) over time,

$$v_x = Kc \, \cos \frac{2\pi z}{\lambda_w} \quad \text{and} \quad v_y = Kc \, \sin \frac{2\pi z}{\lambda_w} \tag{11.7}$$

As the integral yields an additional factor $\lambda_w/(2\pi c)$, the wiggler parameter

$$K = \frac{e}{m} B_w \left( \frac{\lambda_w}{2\pi c} \right) = \frac{e B_w \lambda_w}{2\pi mc} \tag{11.8}$$

As electron velocity is close to the speed of light for our relativistic electrons (FEL works only with relativistic electrons), we must include the effect of relativity defined by Einstein that mass increases with velocity (neglected in Newton's laws). We write for a relativistic electron mass, using equation (10.8),

$$M = \frac{m}{\sqrt{1 - v^2/c^2}} \equiv \gamma m \tag{11.9}$$

From equation (11.9), the mass for a relativistic electron is $\gamma$ times that for a nonrelativistic one. So energy from $E = mc^2$ is also higher by $\gamma$, which from equation (10.8) means that the velocity $v$ for the electron accelerator gun determines the energy of the electrons. Thus, from equation (11.8), $K$ changes to $K/\gamma$ for the relativistic case to give a modified equation (11.7).

$$v_x = \frac{Kc}{\gamma} \, \cos \frac{2\pi z}{\lambda_w} \quad \text{and} \quad v_y = \frac{Kc}{\gamma} \, \sin \frac{2\pi z}{\lambda_w} \tag{11.10}$$

Note that $v_x$ and $v_y$ are small when $v_z$ in equation (11.4) approaches $c$. Using the definition of $\gamma$ from equation (10.8) to compute $v$ and $v_x^2 + v_y^2 = (Kc)^2/\gamma^2$ from equation (11.10), we write

$$\begin{aligned}
v_z^2 &= v^2 - v_x^2 - v_y^2 = c^2 \left( \frac{\gamma^2 - 1}{\gamma^2} \right) - v_x^2 - v_y^2 \\
&= c^2 \left( \frac{\gamma^2 - 1}{\gamma^2} \right) - \frac{K^2 c^2}{\gamma^2} = c^2 \left( 1 - \frac{1 + K^2}{\gamma^2} \right)
\end{aligned} \tag{11.11}$$

For $v \approx c$, $\gamma$ is very large (and $K$ assumed not large), we use the binomial approximation on equation (11.11) to remove the square root for $v_z$:

$$v_z = c \left( 1 - \frac{1 + K^2}{\gamma^2} \right)^{1/2} = c \left( 1 - \frac{1 + K^2}{2\gamma^2} \right) \quad (11.12)$$

The wiggler field **B** is strong enough relative to the electromagnetic field to control the electron trajectory. In some FELs, cryogenic magnets are used. The wiggler field does not exchange energy with the electrons but determines their trajectory.

We write the electron field of the plane wave as

$$\mathbf{E} = E_0 \left[ \hat{\mathbf{x}} \sin \left( \frac{2\pi z}{\lambda} - wt + \phi_0 \right) + \hat{\mathbf{y}} \cos \left( \frac{2\pi z}{\lambda} - wt + \phi_0 \right) \right] \quad (11.13)$$

As the electron has a transverse velocity component and the **E** field is transverse to **v**, the field will exert a force on the electron. As work (energy) $W$ is force times distance, work per second $\dot{W}$ is force $e\mathbf{E}$ times speed **v** (times involves dot product for vectors), using equations (11.13) and (11.10) and $\sin(A + B) = \sin A \cos B + \cos A \sin B$,

$$\dot{W} = e\mathbf{E} \cdot \mathbf{v}$$
$$= eE_0 \frac{Kc}{\gamma} \left[ \cos \frac{2\pi z}{\lambda_{\mathrm{w}}} \sin \left( \frac{2\pi z}{\lambda} - \omega t + \phi_0 \right) + \sin \frac{2\pi z}{\lambda_{\mathrm{w}}} \cos \left( \frac{2\pi z}{\lambda} - \omega t + \phi_0 \right) \right]$$
$$= \frac{eE_0 Kc}{\gamma} \sin \left[ 2\pi \left( \frac{1}{\lambda} + \frac{1}{\lambda_{\mathrm{w}}} \right) z - \omega t + \phi_0 \right]$$
$$\equiv \frac{eE_0 Kc}{\gamma} \sin \phi \quad (11.14)$$

where $\phi$ is defined in equation (11.14) as

$$\phi \equiv 2\pi \left( \frac{1}{\lambda} + \frac{1}{\lambda_{\mathrm{w}}} \right) z - \omega t + \phi_0 \quad (11.15)$$

Here, $\phi$ relates to phase between generated optical or microwave field **E** of wavelength $\lambda$ and electron velocity **v** that depends on wiggler wavelength $\lambda_{\mathrm{w}}$. Note that $\phi = \pi/2$ for in phase and $\phi = 0$ for out of phase.

From Einstein's mass–energy formula,

$$E = Mc^2 = \gamma mc^2 \quad (11.16)$$

where $m$ is the static mass. Therefore, $\gamma$ that depends on $v \approx c$ according to equation (10.8) is proportional to energy. Then, using equation (11.16), we can write equation (11.14) as

$$\dot{W} = \dot{\gamma} mc^2 = \frac{eK_0 Kc}{\gamma} \sin \phi \quad \text{or} \quad \dot{\gamma} = \frac{eK_0 K}{\gamma mc} \sin \phi \quad (11.17)$$

As transverse electron velocity oscillates, see equation (11.10), so does $\phi$ from right-hand side of equation (11.14), which used equation (11.10) in the second line of its development. Hence, from equation (11.17), the rate of change of energy $\dot{W}$ and $\dot{\gamma}$ both oscillate. Using $\dot{z} = v_z$ and equation (11.12) for $v_z$ allows us to write the time derivative of equation (11.15):

$$
\dot{\phi} = 2\pi \left( \frac{1}{\lambda} + \frac{1}{\lambda_w} \right) v_z - w = 2\pi c \left( \frac{1}{\lambda} + \frac{1}{\lambda_w} \right) \left( 1 - \frac{1 + K^2}{2\gamma^2} \right) - \frac{2\pi c}{\lambda}
$$

$$
= \frac{2\pi c}{\lambda_w} \left[ 1 - \left( 1 + \frac{\lambda_w}{\lambda} \right) \frac{1 + K^2}{2\gamma^2} \right] \tag{11.18}
$$

The Compton effect applies to high frequency where the laser wavelength is much shorter than the wiggler wavelength, $\lambda_w/\lambda \gg 1$. So

$$
\dot{\phi} \approx \frac{2\pi c}{\lambda_w} \left( 1 - \frac{\lambda_w}{\lambda} \frac{1 + K^2}{2\gamma^2} \right) \tag{11.19}
$$

In a steady resonant state, the rate of phase change $\dot{\phi} = 0$, which means that the phase difference is constant between the emitted light or microwave electromagnetic radiation and the electron velocity $\mathbf{v}$ that depends on wiggler wavelength $\lambda_w$. From equation (11.16), the resonant electron energy depends on $\gamma$. In the steady state $\gamma \to \gamma_R$, so setting equation (11.19) to zero, the bracket goes to zero and the resonance electron energy $\gamma_R$ (that defines electron velocity for resonance) is

$$
\gamma_R^2 \equiv \frac{\lambda_w}{2\lambda}(1 + K^2) \tag{11.20}
$$

From equation (11.20), the resonant energy depends on the wiggler magnetic field strength $B_w$ and the wiggler wavelength $\lambda_w$ through $K$, equation (11.8). Furthermore, by making $\lambda$ the subject of equation (11.20), we see that the free-electron laser can be made to work at any wavelength $\lambda$ by adjusting $\gamma_R^2$ (the voltage of the electron gun) or by changing the wiggler magnetic strength field $B_w$ or wiggler wavelength $\lambda_w$. This explains the range of frequencies and versatility of the free-electron laser.

## 11.2.2 Electron Bunching for Stimulated Emission in Free-Electron Laser

We observe that the change in phase of the field in traveling a distance equal to the wavelength of the wiggler at resonance can be shown to be $2\pi$ [119]. By substituting $\lambda_w/(2\lambda)(1 + K^2) = 2\gamma_R^2$ from equation (11.20) into equation (11.19) and using equation (11.17) for the first equation, we now have coupled first-order differential equations linking energy and phase:

$$\dot{\gamma} = \frac{eK_0K}{\gamma mc} \sin \phi$$

$$\dot{\phi} = \frac{2\pi c}{\lambda_w} \left(1 - \frac{\gamma_R^2}{\gamma^2}\right) \tag{11.21}$$

These equations describe a single electron in a wiggler field. The electrons gain energy $\dot{\gamma} > 0$ or lose energy $\dot{\gamma} > 0$ depending on the phase $\phi$. Energy gain implies absorption and energy loss stimulated emission. The average value of sin $\theta$ is zero, so there would be no net gain without some other activity, as discussed in Section 10.2. Substituting for $(1 + K^2)$ in equation (11.20) into equation (11.12) and using equation (10.8) gives

$$v_z = c\left(1 - \frac{\lambda \gamma_R^2}{\lambda_w \gamma^2}\right) \quad \text{or} \quad \dot{v}_z = \frac{dv_z}{d\gamma}\dot{\gamma} \approx \frac{2c\lambda \gamma_R^2}{\lambda_w \gamma^3}\dot{\gamma} \tag{11.22}$$

which means electrons are accelerated or decelerated longitudinally depending on whether they are gaining ($\gamma > 0$) or losing ($\gamma < 0$) energy to the plane wave. The bunching of electrons can lead to a net gain rather than absorption. Bunching can be anticipated because when electrons speed up, their mass increases from relativistic considerations, causing them to slow down. The levels of high- and low-energy electrons allow electrons to be stimulated to fall to a lower level for stimulated emission similar to bremsstrahlung in conventional lasers. Gain can be shown to occur when $\gamma > \gamma_R$ and loss if $\gamma < \gamma_R$ [119]. The maximum gain occurs at electron energy [119]:

$$\gamma = \left(1 + \frac{0.2}{N_w}\right)\gamma_R \tag{11.23}$$

where $N_w$ is the number of wiggler periods. For substantial gain, we need a narrow distribution of electron energies:

$$\Delta\gamma \leq \gamma_R/2N_w \tag{11.24}$$

Electron beam sources must deliver a few amperes for substantial gains. Bandwidth of a FEL is estimated from bunch length $l_e$. Laser radiation must be produced in pulses of length $l_c$. So bandwidth $\delta v \approx c/l_e$ or relative bandwidth is

$$\frac{\delta v}{v} = \frac{\lambda}{l_e} \tag{11.25}$$

## 11.3   DESCRIPTION OF HIGH- AND LOW-POWER DEMONSTRATIONS

### 11.3.1   Proposed Airborne Free-Electron Laser

Free-electron lasers were first developed for ships where weight and energy are of less concern than for airborne. By reducing size and weight to below 180,000 lbs, an airborne version is considered feasible: the 747 cargo plane can carry 248,000 lbs. Figure 11.2 shows the possible configuration of a megawatt free-electron laser in a Boeing 747 [170]. The system is designed for multimegawatt power in the extracted optical beam. The injector, cryomodule in the linac (linear accelerator) and beam stop all have superconducting radio frequency cavities. The electron beam is accelerated to 0.5 MeV in the DC electron gun at right. The injector accelerates the beam to 7.5 MeV and the linac accelerates the relativistic beam further to 80–160 MeV, depending on the wavelength of light desired.

The electron beam feeds into the left end of an 18 m straight optical cavity along the lower side of the structure between the optical outcoupling mirror and another mirror at the right end. The wiggler field in the cavity raises the frequency in the optical region of the electromagnetic spectrum. On passing through the wiggler, only about 1% of the energy is converted from the electron beam into light. Therefore, energy recovery linac principles are used [170]; the electron beam energy is recovered by passing the beam with reverse phase up and left back through the cryomodule in the linac. A superconducting radio frequency cavity in the beam stop removes energy provided by the injector and returns the radio frequency power to the injector. A small amount of remaining energy is absorbed at the end of the beam stop.

Studies suggest that 3 MW of electrical energy will be needed to produce 1 MW of laser light. So cooling systems must remove 2 MW of heat. The cooling systems including high-pressure helium compressors and expanders, and heat exchangers, cold box are one of the heaviest items in the free-electron laser. The other heavy components are magnets and magnet chambers.

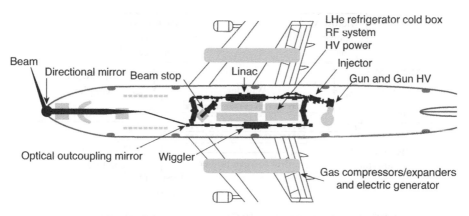

**FIGURE 11.2**   Layout of free-electron laser on an aircraft.

### 11.3.2 Demonstration of Low-Power System for Free-Electron Maser at 8–12 GHz

We describe a laboratory tabletop X-band, 10 GHz free-electron maser [173, 175] for the purpose of illustrating the possibility and complexity of constructing small free-electron lasers and masers. The experimental arrangement (Figure 11.3) shows an electron beam entering a vacuum cavity through a small hole at the left. At the same time, an oscillator seed for generating microwave power is entered through a waveguide at top right. The waveguide ends are blocked, except for the small end holes for electron injection and exit. So the waveguide acts as a cavity that allows a discrete set of microwave resonating frequency oscillation longitudinal modes. The microwave power passes in the backward direction (relative to electron motion), to form a backward wave oscillator (BWO Section 10.1), and interacts with the electron beam in the wiggler to convert energy from the electron beam into microwave energy by stimulated emission (bremsstrahlung). In a backward wave oscillator (BWO) gyrotron [65], the dispersion diagram for the waveguide intersects the fast cyclotron mode for negative propagation constant. The BWO FEM operates in a manner similar to a BWO gyrotron (Figure 10.4) but has the additional wiggler. A coupler at the microwave frequency passes the microwaves to the output at left.

The experimental system (Figure 11.4) [173] is approximately 2 m long and simple to construct and relatively inexpensive (Table 11.1). However, a pulse forming network might be needed to reduce the pulse length and increase the peak power, which is not too high at this time. Reduction of pulse length from microseconds to nanoseconds could possibly result in a peak to 1 MW.

### 11.3.3 Achieving Low Frequencies with FELs

Free-electron lasers have been explored mainly because of their ability to reach much higher frequencies (up to soft X-rays). Extending the free-electron maser frequency

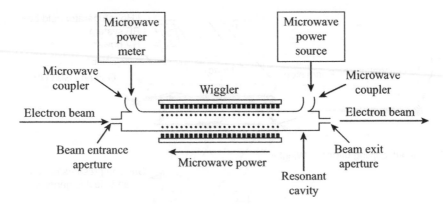

**FIGURE 11.3**   Arrangement for 10 GHz free-electron maser experiment.

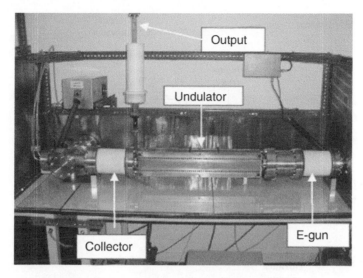

**FIGURE 11.4**    10 GHz free-electron maser experiment.

down to low frequencies such as 10 GHz requires additional work. Typically, this requires operating with low-voltage electron guns (under 100 kV). The electron speed, while still relativistic, is now too low for synchronous operation with the electromagnetic microwave field to generate or amplify the microwave field. So the field should be slowed down. In the experimental system, funded by the European Union, this was accomplished with a 50 kV electron source by periodically loading the waveguide [8] by soldering posts across the cavity at the same period as the wiggler magnets as seen in Figure 11.5 [173]. One meter wavelength appears to correspond with the lowest frequency generated with a free-electron laser [33].

**TABLE 11.1    Specification for Experimental 10 GHz Free-Electron Laser**

| Topic | Item | Value |
|---|---|---|
| Electron beam | Second anode voltage | 80 kV |
| | Cathode voltage | −9 kV |
| | Electron beam current | 200 mA |
| Magnetic wiggler characteristics | On-axis field strength | 0.0325 T (320 G) |
| | Period | 19 mm |
| | Pole gap | 22 mm |
| | Number of periods | 33 |
| | Material | NdFeB |
| Microwave output characteristics | Frequency | 7.9–11.5 GHz |
| | Pulse power | 1 kW |
| | Pulse width | 10–30 μs |
| | Maximum pulse repetition frequency | 10 kHz |

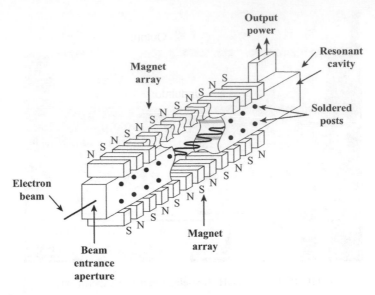

**FIGURE 11.5** Low-frequency modification for 10 GHz free-electron maser experiment.

A possible application of low-frequency free-electron masers is for the penetration of radar into the ground for finding buried objects. However, gyrotrons can also operate at these frequencies, perhaps more easily. Considering the tabletop demonstration unit, the bandwidth of $\Delta f = 4$ GHz gives a range resolution, equation (15.3), $\Delta R = c\tau/2 = c/(2\Delta f)$, of a few centimeters, adequate for finding and identifying small buried objects. How far 5 GHz microwaves penetrate into the ground can be determined by substituting the attenuation coefficient [20]

$$\alpha = \text{Re}\{\gamma\} = \frac{\omega\epsilon''}{2}\left(\frac{\mu}{\epsilon'}\right)^{0.5} = \pi f \frac{\sqrt{\epsilon'}}{c} \tan \delta \qquad (11.26)$$

into the equation for the magnitude of the field as it decays in the earth's interior:

$$u(z) = E_{z=0} \, \exp\{-\alpha z\} \qquad (11.27)$$

to get

$$u(z) = E_{z=0} \, \exp\{-\pi f \frac{\sqrt{\epsilon'}}{c} \tan \delta \, z\} \qquad (11.28)$$

Typical earth values from the *Radar Handbook* [151] are not reliable in the real world because of variability of soil and weather. From equation (11.28), there is an enormous loss in the radiation traveling to and from the buried object at a bandwidth needed for adequate range resolution. While some of this loss can be recovered by using a very high-power megawatt source, this does not change the signal to noise ratio as noise

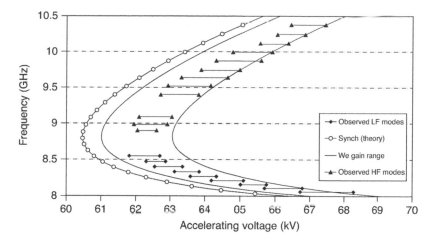

**FIGURE 11.6** Tuning range for operating frequencies of experimental 10 GHz free-electron maser experiment.

is mostly clutter. A synthetic aperture antenna can reduce the noise somewhat by narrowing the beam pattern. Mach–Zehnder interferometric methods (Section 6.1.2) such as the laser Doppler vibrometer will probably work better.

### 11.3.4 Range of Tuning

The output frequency for the experimental unit can be continuously tuned from approximately 8 to 12 GHz, which is a bandwidth of 4 GHz. Figure 11.6 [173] shows how the accelerator (electron gun) voltage (x axis in kV) of the electron gun determines the operating frequency (y axis in GHz). A range is shown for each operating frequency with a line terminated in black triangles. The position on the line is determined by using a seed into the output to select one dominant mode (right hand continuous curve) or the other (left-hand continuous curve). We note that only one frequency in the range 8–10 GHz is obtained at a time, so we do not have the full bandwidth simultaneously as we might in a band-limited carrier frequency.

### 11.3.5 Design of Magnetic Wiggler

The design of a low-cost (<$2,000) permanent magnet wiggler is described in Ref. [174]. It uses only two orientations of magnets rather than the Halbach four-orientation wiggler that performs better but is more expensive. The material used is NdFeB. The end magnets were half the size and overlapped to avoid electrons hitting the walls of the waveguide, a less expensive approach than the normal tapering. Analysis for design and adjustment is discussed in Ref. [174].

**PART III**

---

# APPLICATIONS TO PROTECT AGAINST MILITARY THREATS

---

# CHAPTER 12

# LASER PROTECTION FROM MISSILES

In 1942, the first ballistic missile was launched by Germany that went into the fringes of space at an altitude of 100 km, had a thrust of 56,000 lbs (alcohol and liquid oxygen), a velocity of 3500 mile/h, and a payload of 2200 lbs [124]. Since World War II, the missile has become a dominant weapon of war [140]. A laser beam travels fast enough in air, kilometers in a few microseconds, to track, disable, or destroy an incoming missile before it has time to perform significant evasive maneuvers or to launch countermeasures. The threat of missile attack is increasing with time because more sophisticated platforms for launching missiles are in continuous development: land robot vehicles and unmanned aircraft and submersibles. We note that GPS or radio-controlled vehicles, and to a lesser extent autonomous vehicles, can also be countered by interfering with their free-space communication channels.

In Section 12.1, we introduce the protection from missiles and discuss ways to combat the threat of attack on cities with nuclear multiple warhead intercontinental ballistic missiles (ICBMs). In Section 12.2, we describe the airborne laser program to combat the ICBM threat. In Section 12.3, we discuss protection against attacks by homing missiles on aircraft and moving assets. In Section 12.4, we address targeted attacks by missiles on stationary or slow-moving targets such as ships.

*Military Laser Technology for Defense: Technology for Revolutionizing 21st Century Warfare,*
First Edition. By Alastair D. McAulay.
© 2011 John Wiley & Sons, Inc. Published 2011 by John Wiley & Sons, Inc.

## 12.1    PROTECTING FROM MISSILES AND NUCLEAR-TIPPED ICBMs

### 12.1.1    Introducing Lasers to Protect from Missiles

***12.1.1.1    Laser Requirements and Selection***    The cost of firing a power laser is much less than that for firing an antimissile missile. There is no recoil or after effect. The amount of energy that must be delivered depends on the type of target. In addition, the power and expense per shot may be reduced for training purposes, allowing operators more practice in acquisition, tracking, and firing. Sensors on dummy targets enable computation of accuracy, propagation loss and efficiency.

Lasers in development (Chapters 8 through 11) can produce pulses of sufficient energy to damage soft and hard targets. The energy per unit area, $F$, required to damage a hard target made of metal or steel is on the order of 100,000 J/cm$^2$, while damage to a soft target: flesh, fabrics, plastics, and homing missile sensors needs only on the order of 1000 J/cm$^2$. The energy per unit area, $F$, delivered by a laser to a spot of area $A$ at a target depends on the laser power $P$, the laser pulse length $\Delta t$, and the ratio $L$ of power remaining after propagating to the target:

$$F = P\Delta t \frac{L}{A} \tag{12.1}$$

Requirements for a laser system vary depending on the range to the target, so we separate the discussion into long and short ranges.

***12.1.1.2    Long Range***    Because the missile plume is highly visible on launch, long-range acquisition from an aircraft is feasible during the launch phase. The launch phase is also the best for destroying a missile because while the missile accelerates up through the atmosphere after launch, the missile skin is under maximum stress, at which time the energy required to break or pierce the skin is least. A high-quality laser beam can project energy in milliseconds over distances of tens of kilometers in space or upper atmosphere. Diffraction, causing beam spreading, is reduced by expanding the diameter of the Gaussian beam at its waist to several tens of centimeters using a beam expander (Sections 1.3.2 and 2.1).

It is difficult to destroy missiles or other military targets at long range in the atmosphere because in addition to diffraction, atmospheric turbulent causes the beam to spread and wander (Section 5.2). Spreading and wandering due to atmospheric turbulence can be reduced by adaptive optics in which the anticipated distortion of the beam is measured and a precomputed wave front set adaptively with mirror arrays at the source (Section 5.3). But adaptive optics incurs significant complexity and additional cost. Even with adaptive optics, the loss at long range, hundreds of kilometers, is large. The weakened intensity arriving at the target means that we may have to track a spot on the target for several seconds to provide the required energy for destruction. This allows time for a missile to engage in simple countermeasure maneuvers such as rotation.

At long range, high-power beams can heat the air through which they pass, especially a problem when there is no wind. Heating the path expands the air more toward the center of the path than at the edges because beams are approximately Gaussian (Section 2.1). The expansion on heating reduces the density of the air toward the beam center and hence decreases its density there. The resulting cross sectional refractive index profile for the path is then similar to that for a concave lens that spreads the beam out, reducing efficiency of delivering energy to a small area. The blooming effect is reduced by dithering or spiraling the beam in small circles to avoid heating the same path for too long.

### 12.1.1.3  *Short Range*  At short range, compared to long range, the fast speed of light ($c = 3 \times 10^8$ m/s) gives the missile insufficient time to maneuver, perhaps only microseconds. Atmospheric turbulence and path heating are less significant. At short range, a pulsed laser with short duration, $\Delta t$, may be used (Chapter 9) because geometry is changing fast and less average power is needed. With a pulsed laser, pulse compression can reduce the pulse length while maintaining the energy (Chapter 9). Other effects can occur with very short pulses. For example, with nanosecond lasers, vaporization or ablation can occur in which material bypasses the usual solid to liquid phase and moves directly to a vapor phase. This is used in laser-induced breakdown spectroscopy [120]. The shock waves generated can cause other types of material failures at the atomic level. The plasma generated can reportedly interfere with jet engine intakes. The breakdown of air into a plasma by a laser beam, similar to lightning, produces a conductive path of ions through which microwave frequencies can be focused more finely.

### 12.1.2  Protecting from Nuclear-Tipped ICBMs

The practice of multiple nuclear warheads on a single ICBM [140] makes annihilation of a country feasible. The theory and construction of both nuclear bombs and ICBMs, though difficult, are fairly widely available and efforts can, to some extent, be disguised as activities for developing commercial nuclear power. By the year 2015, we expect more than 10 countries to have multiple nuclear warhead tipped ICBMs that can reach the United States and many other countries. This represents a significant threat.

### 12.1.2.1  *Historical Background*  The philosophy of Mutually Assured Destruction (MAD) was used in the Cold War to protect against nuclear attack. In 1960, the United States stockpiled approximately 19,000 nuclear missiles and warheads [140], more than all known targets and equivalent to 1.4 million Hiroshimas. Delivery was also varied and sophisticated: nuclear tipped missiles could be launched by the Navy from hard-to-detect ballistic missile submarines (Figure 12.1a) [140], nuclear bombs could be dropped with Air Force bombers, and intercontinental ballistic missiles with multiple warheads (Figure 12.1b) [140] could be launched from land-based silos. The USSR was estimated to have had

**(a)**                                                        **(b)**

**FIGURE 12.1** Multiple warhead missiles: (a) multiple missiles in nuclear submarine and (b) multiple warheads in intercontinental ballistic missile.

some 1700 nuclear warheads. Overstocking nuclear weapons is not seen as an effective approach to combating the current nuclear missile threat from rogue nations or groups.

From 1970 until 1983, the Airborne Laser Laboratory (ALL), a modified KC-135A aircraft, was used as a mobile airborne weapon platform [34] for high-power lasers, such as a $CO_2$ laser (Section 8.3.1). In fact, the ALL in 1983 destroyed a sidewinder and cruise missile.

In 1983, the Strategic Defense Initiative (SDI) was announced and immediately dubbed Star Wars [140]. The goal was to completely eliminate the threat from nuclear missiles by preventing any from reaching or exploding over the United States. Given the possibility of hiding a few nuclear warheads among myriads of dummies in multiple warhead missiles, the goal appeared to be elusive. Star Wars provided funding for research into addressing ways to minimize the nuclear missile threat. The USSR was concerned that the United States was circumventing the mutually assured destruction doctrine [140]. Of course, both countries were promoting programs with ambitious goals and extravagant claims to secure funding.

In the late 1980s, the interest shifted to a more specific weapon system the airborne laser (ABL) that was more likely to succeed than the strategic defense initiative. Funding to proceed with development was approved in 1994. The ABL (Figure 12.2) [58] is composed of a Boeing 747 with a powerful chemical oxygen–iodine laser (COIL) (Section 8.3.2) that will, from a steerable nose pod, shoot down ICBMs at hundreds of kilometers range as they leave the lower atmosphere over their launch pads at their most visible and vulnerable.

### 12.1.2.2 *Alternative Approaches to Protecting from Nuclear Tipped ICBMs* The current approach, the ABL program, is described in detail in Section 12.2. In 2010, after 15 years of development, the ABL, for the first time, demonstrated the impressive ability to shoot down an ICBM. However, the proposed system, with many airplanes flying continuously around looking for potential nuclear

**FIGURE 12.2**    Airborne laser to protect from ICBMs.

missile launches, is too complex and costly to handle the geographically spreading threat since the program was started (Section 12.2.6).

One approach is to make a simpler, less costly version. For long ranges, a more powerful laser that avoids toxic chemicals would help. For example, as mentioned in the previous chapters, more powerful free-electron lasers are being developed (Section 11.1) that avoid the storing of dangerous liquid oxygen and iodine for a COIL laser. The current designs of high-power free-electron lasers are large and heavy, making them more useful for shipboard lasers (Section 12.4). Lightweight free-electron lasers are being developed for airborne platforms (Figure 11.2).

A second approach is to ban nuclear testing to curtail future development of more advanced nuclear weapons. At this time, 151 countries have signed and ratified the Comprehensive International Nuclear-Test-Ban Treaty, but nine significant countries have not yet ratified it (Chapter 13). Even if such a treaty is ratified, there is no guarantee that some group will not perform surreptitious testing to develop smaller and more powerful, improved thermonuclear bombs.

In the absence of a viable ICBM missile shield and lacking confidence in an effective comprehensive international test ban (Chapter 13), the ability to continue designing thermonuclear bombs without testing is considered an important deterrent to would-be nuclear assailants. Therefore, several countries, including the United States, are developing extremely powerful inertial confinement lasers that can emulate the temperatures and pressures inside a thermonuclear reaction (Chapter 13). The United States is the leader in this potential new arms race, expecting to demonstrate fusion in next few years. An attractive possibility is that such extremely high-power picosecond inertial confinement lasers will help us understand nuclear fusion sufficiently to produce continuous cost-effective energy by fusion (like the sun) in the next 50 years. Fusion, unlike fission in current nuclear plants, is not expected to pollute future environments with long-lifetime radioactive residual materials and is expected to reduce the possibility of serious accidents from nuclear plants. However, alternative approaches to inertial confinement, such as magnetic confinement, look equally or more promising at this time.

## 12.2   THE AIRBORNE LASER PROGRAM FOR PROTECTING FROM ICBMs

The airborne laser program that started in 1994 involves a Boeing 747 carrying a high-power COIL (Section 8.3.2) that aims to shoot down ICBMs that may be carrying nuclear warheads. The ABL is alerted to a missile launch by a reconnaissance system, such as a satellite, an AWACS airplane, or an advanced radar warning system. The ABL has a suite of infrared (IR) wide-field telescopes and video cameras to detect the plume from a missile launch. A pointing and tracking system tracks the missile and computes an optimum intercept path for the main light beam to destroy the missile while the missile is still under launch stress. The ABL nose turret has a 1.5 m telescope through which it tracks the missile until activating the main beam to destroy the missile. As seen in the airplane (Figure 12.3) [58], six modules of a COIL (Section 8.3.2) are at the rear of the plane, in front of the fuel service system containing stored hydrogen peroxide, chlorine, and oxygen. The beam from the main COIL laser travels in a pipe through a hole in the bulkhead and the crew battle management area to the nose-mounted turret that directs the beam to the target. The bulkhead separates the laser and dangerous chemicals from the crew and optical alignment system.

### 12.2.1   Lasers in Airborne Laser

There are three laser systems in ABL [39] (Figure 12.4). First, the main beam is from a chemical oxygen–iodine laser at wavelength 1.3 $\mu$m with megawatts of power, sustainable for several seconds (Section 8.3.2). The 1.5 m diameter mirror in the rotatable turret focuses the light in a Gaussian beam (Section 2.1) onto the target in the lower atmosphere several hundred kilometers away.

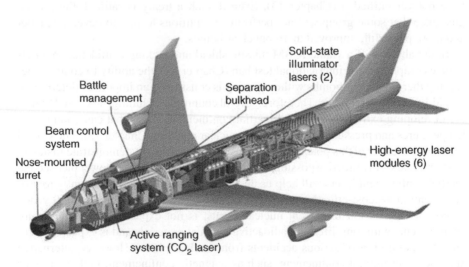

**FIGURE 12.3**   Inside configuration of airborne laser.

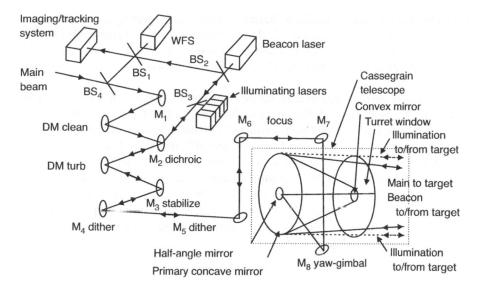

**FIGURE 12.4**   Merging of laser beams in ABL.

The second beam, from a laser diode semiconductor-pumped solid-state laser (Section 8.2) acts as a beacon to illuminate the target missile for the purpose of estimating the turbulence (Section 5.3). In such an adaptive optics system, the beacon estimates the turbulence to compensate for it before transmitting the main beam from the aircraft.

The third beam consists of a set of semiconductor diode pumped solid-state lasers at a wavelength different from the second beam. The illumination lasers that illuminate a large area of the target for the purpose of selecting an aim point on the target on which the main beam will lock for several seconds. The time is long enough to cause the missile skin to destruct or in the case of a liquid fuel missile, the fuel tank to explode.

## 12.2.2   Incorporating Adaptive Optics for Main Beam Cleanup into Airborne Laser

The chemical oxygen–iodine laser beam has inadequate spatial coherence to deliver sufficient power over several hundred kilometers. The devices and the principle of adaptive optics for beam cleanup are described in Section 5.3.

The schematic (Figure 12.4) illustrates how the beam cleanup adaptive optics system (Figure 5.3b and Section 5.3.1.3) merges with the other laser beams as depicted in Ref. [39]. The main beam from the COIL enters at the left and is reflected from a vibration removing fast focusing mirror $M_1$ in an alley of mirrors. The folded arrangement of mirrors reduces the length of the optical path. Because of the high power, lenses or silvered mirrors would burn. So dielectric mirrors are used (Section 6.3), which have alternating high and low dielectric coefficients repeating at period $\Lambda$. This reflects when $\Lambda = \lambda/2$ (Section 6.3.1). The dielectric is made from materials that can withstand the high power, such as a II–VI semiconductor zinc selenide crystal used for the airborne laser window [34]. Materials for high-power lasers are discussed in Section 8.1.5.

The main beam then reflects from the beam cleanup deformable mirror, DM clean (Section 5.3.1.3 and Figure 5.3b), onto a dichroic mirror $M_2$ (Figure 5.4) for merging with the beacon light at a different wavelength. Some light also reflects back into a wave front sensor (WFS) via two beam splitters $BS_2$ and $BS_1$. The measurements from the wave front sensor are used in a computer to set the pistons in the beam cleanup deformable mirror (DM clean) to improve the wave front as described in Section 5.3.1.3. The spatial coherence of the main beam is improved by DM clean on its way to DM turb via mirror $M_2$ dichroic. The cleaned up beam propagates better for focusing to long distances. The optics shown in Figure 12.4 were tested in the laboratory as shown in Figure 12.5 [58].

**FIGURE 12.5** Demonstrating ABL optic's functionality in laboratory.

### 12.2.3 Incorporating Adaptive Optics to Compensate for Atmospheric Turbulence in ABL

The adaptive optics system for compensating for turbulence (Figures 5.4 and Section 5.3.1.4) is incorporated into the optics for the ABL. The beacon laser at top center of Figure 12.4, at 1.06 μm, merges its path with the main beam at the dichroic mirror $M_2$ in which layers of dielectric materials are selected to transmit the beacon laser light at 1.06 μm and reflect the main COIL beam light at 1.315 μm [39]; note that in Figure 5.4 transmission and reflection are reversed to simplify drawing. The two beams do not occur at the same time because the beacon is used to estimate turbulence before the high-power pulse of the main beam is unleashed. The combined beacon and main beam paths reflect from the deformable mirror DM turb used to precompensate for turbulence. The main beam and beacon beam paths pass through fast steering mirrors (FSMs) $M_3$ stabilize for stabilization and $M_4$ dither and $M_5$ dither for dither correction (Section 12.1.1.2).

Then the path passes through two focusing mirrors, $M_6$ and $M_7$, into the rotatable nose turret housing the Cassegrain telescope (Section 1.3.4.1) that acts as a beam expander (Section 1.3.2). In the turret (Section 12.2.5), a yaw-gimbaled mirror, $M_8$ yaw-gimbal, points the beam to a half-angle mirror at the center of the large 1.5 m concave mirror of the Cassegrain telescope (Figure 1.10). The beam reflects to the small convex Cassegrain mirror that spreads the light across the 1.5 m aperture of the large concave mirror. The large mirror focuses the Gaussian beam to produce the smallest spot size possible on the target (Section 2.1). This beam passes through the zinc selenide crystal slice turret window [34] that allows very high-power main beam laser light at 1.3 μm to pass without damage and prevents particles in the air from entering the aircraft.

The beacon and the main beam follow the same paths, but the paths strike the target at almost the same point, but at different times. The time between the beacon pulse and the main beam pulse is too short for turbulence to change noticeably (Chapter 5). The reflected light from the beacon laser is retrieved by the Cassegrain telescope for adaptive optics use. The beacon light returns in the reverse direction along the same path from which it came, to pass through the dichroic mirror $M_2$ dichroic and enter the wave front sensor via beam splitters $BS_2$ and $BS_1$. At this point, the beacon light wave front has distortions from traveling through turbulence to and from the target missile. A computer uses the wave front sensor measurements (WFS) to adaptively adjust the deformable mirror DM turb to remove the distortions in the beacon light due to turbulence. As the main beam and beacon beams travel almost the same paths, the main beam will also be focused correctly through the turbulence (Section 2.1). Thus, the adaptive optics system for compensating for turbulence (Figure 5.4) is incorporated into the airborne laser system (Figure 12.4).

### 12.2.4 Illuminating Lasers for Selecting Target Aim Point

Figure 12.6 illustrates how the illumination lasers illuminate a large area of the target to select and track an aim point. The illumination lasers are a set of semiconductor

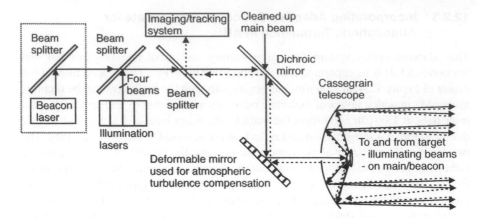

**FIGURE 12.6**   Laser beam for illuminating target for aim point.

diode-pumped solid-state lasers (Section 8.2) of wavelength slightly different from the beacon laser because both beacon and illumination lasers must be reflected by the dichroic mirror but be distinguishable from each other because they are sometimes on at the same time. The beam paths of the illumination lasers are added to the beacon laser beam path at left in Figure 12.6. The illumination and beacon laser beams merge with the beam path of the main beam at the dichroic mirror. In Figure 12.6, for the dichroic mirror selected, merging arises because the illuminating and beacon lasers reflect from the dichroic mirror while the cleaned up main beam is transmitted.

The cleanup of the main beam is described in Section 12.2.2 and compensation for turbulence using the beacon beam in Section 12.2.3. Each of the illuminating lasers focuses to a different point on the large mirror of the Cassegrain telescope. So there is a separate beam for each illuminating laser. The returning light from the reflection of the illuminating lasers from the target (dashed lines in Figure 12.6) is used to form an image with the imaging and tracking system. As the main pulse, for long distances, may have to stay on the same point of the target for several seconds, the telescope must steer to track the spot precisely on the missile. Unlike the beacon laser, the illuminating lasers can remain on during the main beam pulse.

### 12.2.4.1   *Incorporating Illuminating Lasers into Airborne Laser*   Figure 12.4 shows the set of illuminating lasers merging with the beacon laser beam of almost the same wavelength via beam splitter $BS_3$, as suggested in Figure 12.6. The returning light from the target passes all the way back past beam splitters $BS_3$, $BS_2$, and $BS_1$ to the imaging and tracking system. Note that in Figure 12.6, for convenience, the imaging/tracking system is drawn closer to the Cassegrain telescope than the illumination lasers, while in Figure 12.4 (from Ref. [39]), the imaging/tracking system is drawn farther from the Cassegrain telescope than the illumination lasers. The different order is not considered significant.

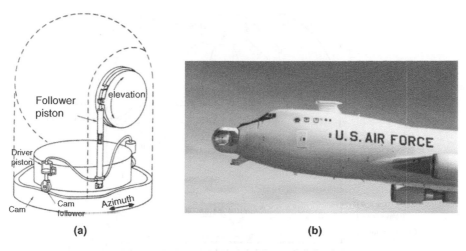

**FIGURE 12.7** Rotating turret of airborne laser: (a) mechanism for pointing and tracking and (b) looking sideways.

## 12.2.5 Nose Turret

The mechanism for rotating the nose for the airborne laser in azimuth and elevation is shown in Figure 12.7a [34]. In Figure 12.7b [58], the nose turret in the airborne laser plane is looking sideways toward the photographer. A block diagram showing layout of the optics in the pointing and tracking system is shown in Figure 12.8 (see Ref. [34] for details).

## 12.2.6 Challenges Encountered In the ABL Program

Tests to shoot down intercontinental ballistic missiles have recently succeeded. Successfully completing such a complicated system with so many diverse systems that have to be developed [34] is an extraordinary accomplishment that will be immensely valuable in future laser weapon systems. For this application, several potential problems were identified.

1. *Coverage Requires Fast Long-Distance Acquisition*: To detect launches of missiles from all potentially dangerous areas of the world with a limited number of planes places an extraordinary requirement on the airplane acquisition system in terms of speed and distance.

2. *Turbulence Compensation Requires Complicated and Expensive Adaptive Optics*: In the upper atmosphere, turbulence and air path heating cause the main destruction beam to spread and wander, necessitating a complicated system of compensating adaptive optics that is expensive and reduces performance.

3. *Maintaining Beam on Target Spot Requires Accurate Long-Distance Tracking*: The weakened and wandering laser spot at the target, which is at some distance

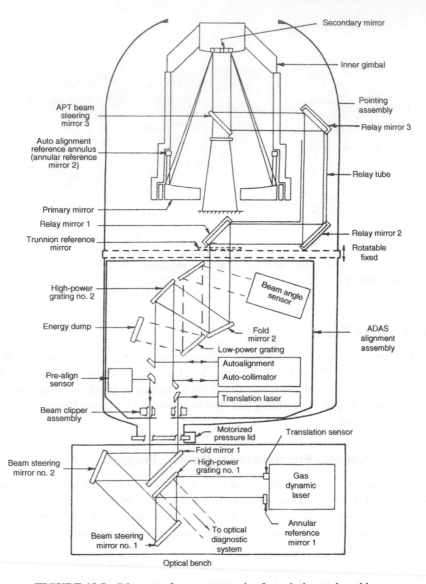

**FIGURE 12.8**  Diagram of nose turret optics for pointing and tracking.

from the airplane during the launch phase, requires that the beam be held on the same area of the target for several seconds. The relatively long time that the beam must be held on the same target area requires excellent and accurate tracking capability.

4. *Time Lags Allow Missile to Maneuver*: The longer is the required tracking with the main beam because of a weakened beam, the more time the missile has to execute countermeasure maneuvers such as twisting on-axis.

5. *Sensitivity to Bad Weather*: Bad weather leading to excessive turbulence and beam attenuation can further reduce the performance of the system.

6. *Missile Coating*: The missile can be coated with mirror or dielectric mirrors to reflect frequencies of interest to reduce the ABL performance.

7. *Carrying Dangerous Chemicals on the Plane*: Toxic and dangerous chemicals such as basic hydrogen peroxide, chlorine, and iodine create a hazard for the crew, especially in a battle situation.

8. *Optical System Sensitivity*: The 1.5 m diameter mirror that points the beam at the target requires pointing accuracy and accurate alignment, making the optical system highly sensitive to airplane maneuvers or vibration.

9. *Speed Vulnerability of ABL*: The airplane is large and slow. It presents an easy target to enemy weapons.

10. *Reload Time Vulnerability of ABL*: The heat generated by firing the laser must be dissipated before another shot is taken. This causes the plane to remain in the same location for a noticeable time.

11. *Eliminating Laser Exhaust Signature Requires Expensive Cryoabsorption System*: In order not to provide a large infrared signature and announce the presence of a high-power COIL laser (Section 8.3.2), ABL needs to have a complicated expensive cryoabsorption vacuum pump to absorb waste gases (Section 8.3.2.3).

12. *New More Powerful Lasers May Make the System Obsolete Before It Can Be Deployed*: Nonchemical lasers such as the free electron laser are being developed (Section 11.1). These avoid the storing of dangerous chemicals, such as hydrogen peroxide, liquid oxygen, and iodine, used with the current COIL laser. They are also tunable and can operate at almost any frequency.

The ABL program showed that each of the above problems could be addressed. However, continuously flying such a complex system around an increasing number of nuclear missile threat areas of the world is unlikely to prove sufficiently cost-effective and reliable. While the future of this 15 year-old program seems to be uncertain, the results with be of immense value for future military laser systems.

### 12.2.7 Modeling Adaptive Optics and Tracking for Airborne Laser

In Section 5.4.1, we described a numerical approach to modeling turbulence in propagating a beam through atmospheric turbulence over distance. As turbulence varies over the path from start to end, the method uses a layered model. Fresnel diffraction is used to propagate from one layer to the next while ignoring turbulence. At each layer interface, a phase screen changes the phase to account for the turbulence encountered since the last phase screen. The computation for the computationally demanding phase screen uses a Kolmogorov statistical model and is described in Section 5.4.2 and 5.4.3. Such phase screens are used in an adaptive optics and tracking testbed in the Advanced Concepts Laboratory (ACL) at MIT Lincoln Laboratories [37].

**12.2.7.1  *Reason for Airborne Laser Testbed*** To design such a complex
dynamic system as ABL, a testbed was needed.

1. The testbed provides records of time history of intensities and gradients,
   reconstructed phases, tracker tilts, main beam pointing, jitter, and beam intensity at target. Not only do these records enable system optimization but they
   are also available for comparison with the measurements for a physical system
   to ensure everything is operating as planned.
2. The testbed results can be cross-checked with wave optics simulations and with
   a physical system.
3. Many test scenarios may be tested with reproducible turbulence for selecting the
   best approach and optimizing the configuration. In the real world, turbulence
   changes from one flight to another reducing the validity of comparisons, not to
   mention the longer time required to perform test flights. Anticipated effects of
   wind and turbulence levels are more easily determined with a simulator.

**12.2.7.2  *Description of ABL Adaptive Optics and Tracking Testbed***
Figure 12.9 shows the three sections of the testbed [37]: a transmit/receive bench for
adaptive optics to compensate for turbulence (sections 5.3 and 12.2.3), beam propagation through the atmosphere with layered rotating phase screens (Section 5.4.2)
and target bench with model missile. The transmit/receive bench and the target bench
can be moved in 3D by actuators for tracking and pointing, in the same manner as
seats in flight simulators and video games. Wavelengths are scaled down by about
half relative to actual wavelengths to reduce size as explained next.

The *transmit/receive bench* includes a wave front sensor (Section 5.3.1.1), a 241
actuator deformable mirror (Section 5.3.1.2), image tracker and fast steering mirror,
adaptive optics and tracking illuminators (Section 12.2.4), and a scoring beam. The
scoring beam is a surrogate for the main beam from a high-energy laser (HEL), in
this case a COIL (Section 8.3.2) at 1.315 μm. The scoring beam is powered by a
helium–neon (HeNe) laser at 0.6328 μm, approximately half of the wavelength of the
COIL.

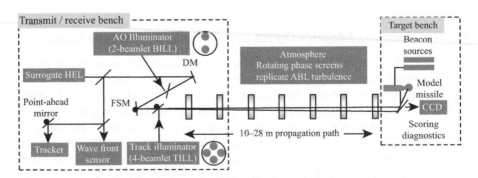

**FIGURE 12.9** Diagram of adaptive optics and tracking testbed.

The adaptive optic (Figure 12.9) AO illuminator performs the role of the *beacon illuminator* (BILL) described in Section 12.2.3 for adaptively compensating for turbulence. In the testbed, the lines of an argon-ion laser provide the BILL for compensating for turbulence at 0.514 μm and the tracking illuminator (TILL) at 0.488 μm. The actual BILL and TILL are generated by Nd:YAG lasers (Section 8.2) of slightly different wavelengths around 1.06 μm. The TILL has four beams as shown in Figure 12.9 that are each slightly divergent to flood the target model missile with light.

The BILL or beacon laser spot on the target returns light to the transmit/receive bench where the WFS measures the wave front over the cross section to set the actuators in the DM adaptively so that the aberrations due to turbulence are removed. The wave front is shaped so that the Gaussian main beam and beacon beams are focused by the Cassegrain telescope (Section 1.3.4) to a tiny spot on the target model missile. The return light from the TILL provides the tilt of the beam that is used to control the fast steering mirror (FSM).

The *propagation path L* can be varied from 10 to 28 m for the specific application scenario to meet the Fresnel index [163] $F = D^2/(\lambda L)$, that is, brightness at the Gaussian image point in the absence of aberrations. $F$ is dependent on aperture $D$ and inversely dependent on distance $L$ and wavelength $\lambda$. Seven stations along the path (Figure 12.9) allow two phase screens at each station. Each phase screen models the effects of turbulence over the distance from the last phase screen. Because of the lengthy computation for phase screens, phase screens are precomputed as described in Section 5.4.2 and 5.4.3. Each phase screen is etched finely around a 6 in. diameter fused silica disk to provide phase inhomogeneities to model accumulated turbulence effects since the last phase screen. At the altitude of the airborne laser, atmospheric turbulence is weak, so 14 phase screens are adequate for modeling over the distance to the target. The strength of the turbulence is determined by the depth of the etch. Each disk has 1 in. wide annular rings of varying turbulence strength. The scoring beam, representing the main beam, passes through the center of an annulus ring. Rotating the disk causes the phase screen to shift, simulating wind across the beam path. A strong single direction wind will reduce blooming due to heating of the air and will have consistent effect on the adaptive optics system.

For a specific engagement scenario, the phase screens are selected to match a number of parameters [4, 37, 144, 163].

1. The turbulence profile $C_n^2(z)$, equation (5.31), is a function of strength of turbulence along the path.
2. The parameter $r_0$ is the atmospheric coherence length or *seeing eye size* above which diameter no further resolution is obtained with a telescope because of turbulence. Adaptive optics is used with new terrestrial telescopes to increase resolution beyond this limit in an effort to compete with space telescopes.
3. The Rytov variance $\sigma_R^2$ is a measure of the strength of scintillation that describes fluctuations in received irradiance resulting from propagation through turbulence; for example, stars twinkle for temporal variation and speckle for spatial variation. The Rytov approximation for weak turbulence involves

multiplication of perturbation terms [4]. Phase screens developed for the airborne laser simulation cover the range of atmospheres $0 < \sigma_R^2 < 0.7$.

4. In adaptive optics, to remove turbulence with a beacon beam, the wave front from the beacon beam is used to set the deformable mirror so as to remove effects of turbulence from the main beam prior to launching it. Therefore, the effectiveness of the adaptive optics turbulence compensation depends on staying within the isoplanatic angle, $\theta_0/(\lambda/D)$, which is the largest angle at the wave front sensor for which the main beam and the beacon beam are reasonably similar. In this case, the difference in frequency between main beam and BILL, or chromatic isoplanatic angle, matters.

5. The adaptive optic electronic or computer controller controls the feedback loop in an adaptive optics system [144]. The controllers task is to take a vector $\mathbf{m}_s$ of readings from the wavefront sensor WFS and compute a vector of settings $\mathbf{c}$ for the actuators in the deformable mirror DM. Note that both WFS and DM are 2D and we simplify the explanation by assuming a linear relation and entering the 2D values into 1D vectors. The controller performs the computation

$$\mathbf{c} = \mathbf{M}\mathbf{m}_s \tag{12.2}$$

where matrix $\mathbf{M}$ is computed to minimize the difference between slopes in the wave front sensor and deformable mirror using commonly used *maximum a posterior* and *least squares* estimation [161]. An alternative more complicated approach minimizes the aperture average mean square residual phase error using wave front statistics and WFS noise statistics. The feedback loops have bandwidth $f_B$ for compensating for turbulence and $f_T$, the Tyler frequency, for compensating for tilt due to delays in the loops. A characteristic frequency for atmospheric turbulence is the Greenwood frequency $f_G$. The orientation and the velocity of each screen are selected to reproduce the ratios $f_B/f_G$ and $f_B/f_T$.

The *target bench* carries a centroid monitor for estimating far-field jitter and a CCD camera to measure the intensity profile of the scoring beam across the target. The Strehl ratio relates intensity on-axis for the turbulent aberrated beam to the intensity on-axis of an unaberrated beam. The Strehl ratio is measured by taking short and long exposure readings of the scoring beam, peak to total integrated intensity compared to that for a diffraction-limited beam. Illumination back from the target is obtained in the target bench by reflection from a model missile or with a beacon point source illuminator or an extended width beacon at wavelengths 0.488 and 0.514 μm. The latter BILL and TILL sources are passed through a cut out of the model missile and then reflected from a rotating retroreflective disk to remove scintillation before returning light to the transmit/receive bench through the propagation phase screen path.

**12.2.7.3 *Experimental Testbed Simulation Results*** One of the complications in the ABL is that the optical tracking system and the adaptive optics turbulence compensation share the same optical paths at the same time. Three sets of experiments of increasing difficulty with both systems in operation were designed to allow separate assessment and optimization of each system while both are present.

The first set of experiments used a point source for both tracking and adaptive optics for turbulence. In the second set, a point source was used for tracking, but a more realistic extended source over an adjustable area to represent reflection from a missile was used for adaptive optics turbulence compensation. In the third set of experiments, the model missile was used to provide still more accurate modeling. The results show reproducibility of experiments, critical for comparing different configurations to optimize the overall system. Each system performance was assessed independently. The results agreed with lengthy computer computations. The results are presented in detail in Ref. [37] The simulation clearly played a significant role in designing the final optical tracking and adaptive optics turbulence compensation systems.

## 12.3 PROTECTING FROM HOMING MISSILES

### 12.3.1 Threat to Aircraft from Homing Missiles

Single person shoulder fired missile launchers were introduced in the 1950s and are now produced in 25 countries. They sell for around $10,000 but can be purchased between $100, for old models, and $250,000, for the latest models. They are extremely effective against slow targets with limited maneuverability, including helicopters, transport planes, and almost any aircraft during take-off and landing. A modern IR seeking missile may sense more than one wavelength, for example 3–5 and 8–13 μm. The missile performs image processing such as contour imaging that makes commonly used flares and point countermeasures ineffective. More recently, the advent of lasers in the countermeasure systems has allowed aircraft to be reliably protected. However, the countermeasures should continue to evolve with the missiles to maintain this level of protection. The cost of protection is orders of magnitude less than the value of a military plane, including its value by remaining in service. For a civilian plane flying in nondangerous places, the risk of downing by a missile is so small that it makes it difficult to justify up to $1 million per plane for protection from missiles. However, the public fear of flying can be escalated for commercial purposes as people might pay more to fly on a protected plane and costs will decrease with volume.

***12.3.1.1 Missile Guidance Mechanisms*** The closed-loop guidance system is the most vulnerable part of the missile guidance because a small error in steering can cause the missile to miss the target. Steering involves an antenna array: for infrared optics, this may be a CCD. Missile guidance has been well studied as a time-varying problem of nonlinear parameter estimation of the direction of arrival (DOA) of a plane wave incident at an angle $\theta$ with the normal on the array (Chapter 14). Reference [162] derives optimum processors. As indicated in Chapters 1 and 14, finding the direction of arrival of a plane wave source with an array is equivalent to finding the component of its wave number **k** (or propagation constant) along the array from the following equation:

$$k \sin \theta = \frac{2\pi}{\lambda} \sin \theta \tag{12.3}$$

Computations in beam space, $\theta$–$\mathbf{r}$, are often simpler than those in $\mathbf{k}$–$\mathbf{r}$. A missile must perform computations rapidly in a fast changing environment so that a cost–accuracy–time trade-off is critical. According to Ref. [162], linear prediction algorithms such as Levinson–Durbin [83] perform less well than MUSIC and ESPRIT. For best effect, the developers of countermeasures against homing missile must understand the typical guidance systems used in homing missiles.

### 12.3.2 Overview of On-Aircraft Laser Countermeasure System

#### *12.3.2.1 Goal of Countermeasure System*   The countermeasure system should accurately track the missile, determine its threat level, and then if the threat is serious, interfere with the missile guidance tracking control loop, so that the missile will lose its lock and miss the aircraft. Concurrently, the laser beam from the countermeasure will dazzle and at closer range damage the missile IR sensor.

#### *12.3.2.2 Overview of Operation and Parts*   Figure 12.10 shows a homing missile tracking the engines of a transport plane with a wide field of view as marked [167] (U.S. patent 5,600,434). The airplane countermeasure system for homing missiles, located on the underbelly behind the wing, responds by sending a laser main or defense beam in the direction of the missile to foil it. A typical aircraft countermeasure system may include a wide-angle IR tracker, an interface to the aircraft radars and sensors, a passive IR camera pointing and tracking subsystem, a powerful pulsed main defense beam whose pulse rate and wavelength can be varied to eliminate the missile threat, and a continuous laser beam for active pointing and tracking,

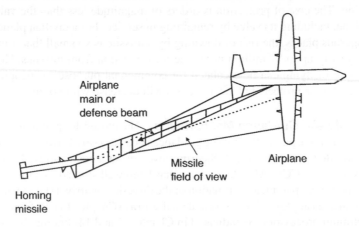

**FIGURE 12.10**   Homing missile tracking transport plane.

vibration elimination, adaptive turbulence compensation, and selection of main beam aim point. A shutter allows the reflection from the missile for the continuous wave to be viewed only while the main beam pulse is off. This shutter provides a time gate to significantly reduce clutter.

### 12.3.2.3 *Description of Countermeasure Pods*   The *man portable air defense system program* (MANPADS) was completed for the Department of Homeland Security (DHS) in 2010. In this program [62], countermeasure homing pods on an airplane protect the airplane from homing missiles such as those fired from the shoulder. The pods may be streamlined units that are built into the plane with a turret sticking out, such as the BAE Jeteye pod [59] (Figure 12.11). Figure 12.12 shows such a BAE

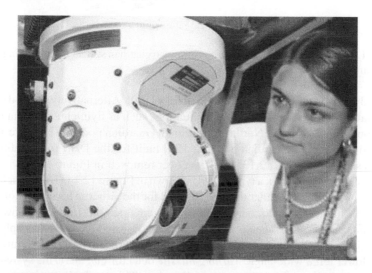

**FIGURE 12.11**   Turret of built-in BAE Jeteye countermeasure system.

**FIGURE 12.12**   Testing BAE Jeteye countermeasure system on an American Airlines jet.

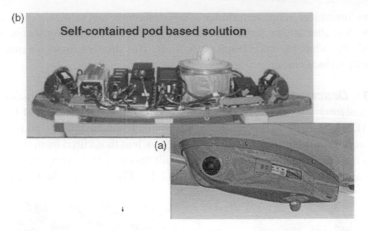

**FIGURE 12.13** Countermeasure Northrop Grumman guardian system for retrofitting to military aircraft: (a) pod and (b) inside of pod.

Jeteye pod on an American Airlines commercial jet [59]. Alternatively, the pod may be an external add-on that can be added for situations such as flying humanitarian aid to a dangerous region, as in the Northrop Grumman Guardian pod [128] (Figure 12.13). A Northrop Grumman Guardian pod [62, 128] built for the DHS for retrofitting to aircraft is shown in Figure 12.13a, with its cover removed in Figure 12.13b.

A diagram of the inside of a typical pod is shown in Figure 12.14 [167]. Much of the functionality required is similar to that described for the airborne laser (Section 12.2). However, the homing missile countermeasure system is simpler, has lower power, and is more compact because range is a thousand times less and the beam has only to interfere with the missile IR sensors and not burn through the missile skin. Figure 12.14 shows at lower left a wide-angle *warning missile* detector that alerts the system to an approaching missile and its approximate direction. This works in conjunction with aircraft radars and avionics. The electronic countermeasure device (ECM) shown in Figure 12.14 is also alerted, if warranted, to jam the radar of a guided remote control

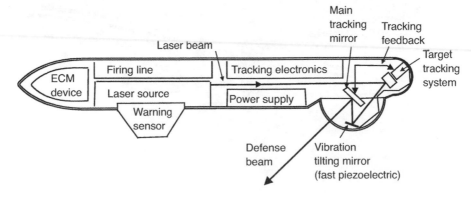

**FIGURE 12.14** Diagram of inside of a countermeasure pod used against homing missiles.

missile. If the missile is deemed a threat, the passive and active tracking systems are activated. For an underbody system, the transparent dome at right sees a complete hemisphere below. More pods may be required for protection of large planes from several directions.

### 12.3.3 Operation of Countermeasure Subsystems

***12.3.3.1 Passive Missile Tracker*** Because of its high speed, a homing missile gets hot in front of the dome of its homing head and at its wing tips. An IR camera in the aircraft pod optical tracking head observes these hot spots and tilts the main tracking mirrors to lock onto these with a feedback loop (Figure 12.14) [167]. Once a lock is achieved, the firing line is tripped to activate the power lasers. The locked tracking directs all laser beams from tracker to missile.

***12.3.3.2 Pulsed Main Power Beam*** A laser diode-pumped Nd:YAG solid-state laser in the aircraft pod (Section 8.2) is pulsed at the reticule rate used by the electronics in the missile to read the missile CCD. This will cause incorrect estimates of the aircraft direction and interfere with the guidance control so as to break its lock. Even without this effect, the laser from the aircraft will cause dazzle and glare in the missiles sensors and at close range can damage the sensors. The pulsed power laser is not pointed at a hot spot used for tracking because it has high reflectivity. Instead, a camera and electronics select an aim point on the missile head that is less reflective and hence more prone to absorbing the laser energy, as well as more critical to the correct functioning of the missile, for example, the sensor area. Several lasers of different wavelengths may be used as shown in Figure 12.15 [167], which includes

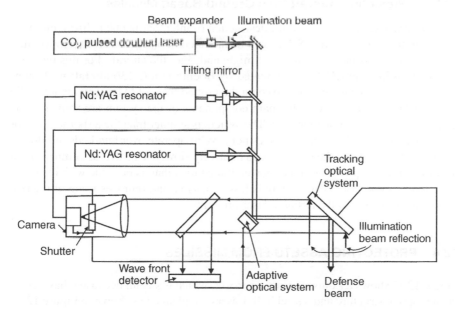

**FIGURE 12.15**   Lasers and adaptive optics system to compensate for turbulence.

a doubled carbon dioxide laser (Section 8.3.1) at a missile light wavelength of 5 μm (a frequency that has low attenuation through the atmosphere, Figure 16.1), a ND:YAG laser (Section 8.2) at 1.06 μm doubled to 0.53 (Section 8.2.2), and a second Nd:YAG frequency tripled to 0.25 μm for ultraviolet. A countermeasure system can also include an optical parametric amplifier (OPO) consisting of a nonlinear crystal to tune a laser over a wider range for countering more sophisticated missiles.

### 12.3.3.3 Low-Power Laser for Adaptive Optics

Atmospheric turbulence causes the relatively narrow beams to diverge and wander around because of rotating eddies in the air due to atmospheric turbulence, to which the motions of the missile and aircraft contribute. These eddies have less density at the edges and act as lenses of varying diameter and orientation (Chapter 5). The system is greatly enhanced by using a continuous wave beam with an adaptive optics system shown in the lower part of Figure 12.15. The returning illumination beam reflection at lower right from the missile into the tracking optical system, is directed by the tracking optical system mirror to a wave front detector (Section 5.3.1.1). A shutter is used to allow light to pass to a CCD camera only when the main power light is zero between pulses. The location of the returned light on the camera is used to move the adaptive optics system mirror to keep the focus on the aim spot of the array.

The shutter also provides a time window that significantly reduces clutter. As the missile target starts at a point at far distance, light reflecting from other places will produce clutter that is difficult to distinguish from the target. The shutter time window blocks those reflections that are not in the same time window.

### 12.3.4 Protecting Aircraft from Ground-Based Missiles

The airborne laser was designed for destruction of intercontinental ballistic missiles. However, the COIL (Section 8.3.2), can be mounted in other aircraft to remove missile carrying assets on the ground that might endanger the aircraft. For this purpose, a downward looking COIL was mounted by Boeing in a C-130 aircraft to demonstrate the ability to damage a Humvee on the ground that could be carrying a missile launcher [1] (Figure 12.16). The main components of the system are shown in the figure and described in Section 12.2. The system is simpler than the airborne laser for shooting down missiles because it does not have to operate over hundreds of kilometers from the target and the target moves relatively slowly. This avoids compensation for turbulence and provides higher power at the target than is possible with the nose-mounted airborne laser. This system is less expensive and removes some of the risk factors associated with the airborne laser program.

## 12.4 PROTECTING ASSETS FROM MISSILES

Figure 12.17 shows a Zumwalt class destroyer, which is stealth because it has a composite topside structure and a steel hull. Lasers are planned as shown in Figure 12.17 to shoot down aircraft, missiles, other ships, and shore targets. Ships may have both

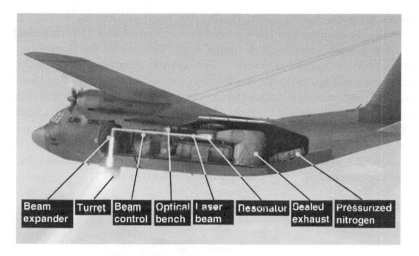

| Beam expander | Turret | Beam control | Optical bench | Laser beam | Resonator | Sealed exhaust | Pressurized nitrogen |

**FIGURE 12.16**   Flying COIL laser on a C-130.

short-range and long-range lasers to shoot down missiles. For short range, because of the high speed of light, relatively low-power lasers can be used to shoot down missiles that have managed to come close to the ship. These lasers act similar to the homing missile countermeasures for aircraft (Section 12.3).

Lasers used in place of heavy artillery avoid carrying arsenals of high explosives that have to be protected against attack. The free-electron laser (Section 11.1) is a candidate for the lasers on a ship because weight and pump power are much less of a problem than on an aircraft. Unlike the COIL used in the airborne laser, the free-electron laser does not require toxic chemicals on board. It requires only electric

**FIGURE 12.17**   New stealth ship with lasers.

generators for powering the lasers. Pulsing the lasers reduces the average power (Chapter 9). The free-electron laser is tunable over more than a 15% range and can be built for a much wider range of frequencies. This allows the frequency to be chosen for good atmospheric propagation. For example, 3 mm electromagnetic waves, between optics and microwaves, pass through fog, mild rain, drizzle, and sea haze (Figure 16.1) and are also hazardous to electronic systems on missiles and satellites (Section 17.4). The problem of sea haze is discussed in the introduction to Chapter 14 and (Section 16.3.4) [145]. Laser beams travel straight and cannot shoot over the horizon as conventional artillery or missiles, but high-energy lasers have the power to shoot satellites in earth orbit or destroy them by punching holes in their solar cells and map satellites and debris in lower earth orbit (Section 16.3.1).

# CHAPTER 13

# LASER TO ADDRESS THREAT OF NEW NUCLEAR WEAPONS

Testing thermonuclear weapons for the military is harmful for the environment. So in 1996, 151 countries signed and ratified the Comprehensive Nuclear-Test-Ban Treaty. Subsequently, international pressure is on the significant countries who signed the treaty but have not ratified it, for example, China, Egypt, Indonesia, Iran, Israel, and the United States, and those who have not yet signed the treaty, for example, India, North Korea, and Pakistan. Without the ability to perform nuclear tests, the legal design and development of novel thermonuclear (fusion) bombs would languish. Unfortunately, a rogue country or group might continue surreptitious testing to develop nuclear bombs to everyone's disadvantage.

In Section 13.1, we discuss the justification to construct a super laser that can generate the conditions inside a thermonuclear bomb in the coming age of testban treaties. In Section 13.2, we describe the structure and operation of the National Infrastructure laser.

## 13.1 LASER SOLUTION TO NUCLEAR WEAPONS THREAT

### 13.1.1 Main Purpose of U.S. and International Efforts

A solution to this threat for the United States was proposed: create an extreme laser that can emulate the temperature and pressure conditions at the center of a thermonuclear

*Military Laser Technology for Defense: Technology for Revolutionizing 21st Century Warfare*, First Edition. By Alastair D. McAulay.
© 2011 John Wiley & Sons, Inc. Published 2011 by John Wiley & Sons, Inc.

bomb. The plan is to use a solid-state laser like that we described in Chapter 8, Section 8.2. Two countries are making substantial progress in building the most powerful lasers in the world for this purpose, the United States and France, both of whom have collaborated on massive nuclear research lasers in the past. The National Ignition Facility (NIF) at the Lawrence Livermore National Laboratories [127] is near completion; Lawrence Livermore is the home of nuclear bomb development in the United States. The French program at Bordeaux, called Megajoule [12, 117, 143], has similar capability to NIF and is planned for completion in 2014, also targeted to the nuclear weapons industry. Because of strong collaboration between the United States and France, the facilities and laser design are similar. Both use solid-state lasers of neodymium in glass, which handles much higher power than neodymium in YAG (Section 8.2). The NIF laser is expected to allow the United States to continue to develop, without atmospheric testing, thermonuclear bombs, a critical capability for the military.

Success in either program could stimulate an international race that few countries can afford; the French have suggested that countries with nuclear aspirations have some access to these lasers and/or the research to feel less threatened, but this could strengthen the club of nuclear countries. A list of other known competitive extreme laser projects include SG-III in China, GXII in Japan, OMEGA in Rochester in the United States, and HiPER in the United Kingdom.

### 13.1.2    Benefits of Massive Laser Project

First, the project maintains, develops, and trains experts in nuclear weapons and high-power lasers that are critical should an emergency occur. In this regard, it supports higher education that continues to attract brilliant scientists and engineers from across the globe. Second, it allows scientists and engineers to better understand thermonuclear fusion, the hope for future clean energy that does not create waste that is dangerous to man for hundreds or thousands of years and that no one is happy to have stored in their backyard where it could be released accidentally in an earthquake or other natural disaster. Nuclear fusion (combining) contrasts nuclear fission (splitting) that is used today. At this time, for nuclear fusion, magnetic confinement looks more suitable for power generation than inertial confinement discussed here.

### 13.1.3    About the NIF Laser

The NIF laser was constructed from 1997 to the present and is near completion [127]. The plan is to safely focus a very short laser pulse of a few picoseconds ($10^{-12}$) onto a very small pellet (Section 13.2.1) consisting of a few milligrams of a deuterium–tritium mixture. The terawatt ($10^{15}$) peak power laser pulse will raise the temperature, pressure, and density enough to achieve a controlled nuclear fusion chain reaction (ignition) by Inertial Confinement Fusion (ICF), in a manner similar to the sun. As a result of fusion, the mass $m$ is converted into energy $E$ according to Einstein's famous mass–energy equation $E = mc^2$ [15]. The squaring of the speed of light $c = 3 \times 10^8$

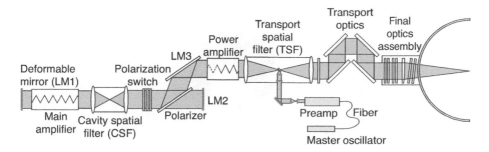

**FIGURE 13.1**   One of the 48 similar beamlines in the NIF.

means that a great deal of energy is obtained from a very small mass, which is why thermonuclear bombs are so powerful and the sun lasts so long.

## 13.2   DESCRIPTION OF NATIONAL INFRASTRUCTURE LASER

### 13.2.1   Structure of the NIF Laser

The NIF laser starts by generating a short pulse with a master ytterbium doped (1053 nm) fiber laser oscillator. This signal is distributed to 48 identical beamlines via an optical fiber; a typical beamline [127] is shown in Figure 13.1. The pulse from the master oscillator at lower center enters a preamplifier.

All the preamplifiers and amplifiers use flashlamp-driven Nd:glass disks, a material that withstands higher powers than Nd:YAG. Rods similar to much larger versions of those used in solid-state Nd:YAG (Section 8.2) were used in earlier inertial confinement lasers: an example is shown of a Nd:glass rod used in France in 1970 (Figure 13.2) [143]. The maximum power is limited by that for which the glass rod fractures or the beam becomes overly distorted. The power of an amplifier can be increased 10 times by replacing a rod by a series of disks, angled alternately forward and backward at the Brewster angle, to eliminate reflections from their surfaces (Figure 13.3) [36]. Disks are thin, avoiding the uneven heating due to different flashlamp penetration distances into the rod that causes rods to fracture. As a result, disks can be pumped with up to nine times the power of rods [36]. The expanded beam diameter also causes much less glass deterioration at high intensities. The zigzag lines in the amplifiers in Figure 13.1 represent angled disks. One of the disks that replaced the rod in Figure 13.2 in the French laser is shown in Figure 13.4 [143].

The light entering the center of Figure 13.1 is reflected to the left into a power amplifier and then down by mirror LM3 through a polarization switch that directs it into a spatial filter to restore spatial coherence (Section 1.3.6). After passing the main amplifier, the light is reflected by mirror LM1 that is deformable to correct for aberrations. Because the flashlamp incoherent light duration is longer than the main laser beam pulse duration, the amount of light coupling from flashlamp to pulse is enhanced by allowing light to bounce back and forth between LM2 and LM1 until the

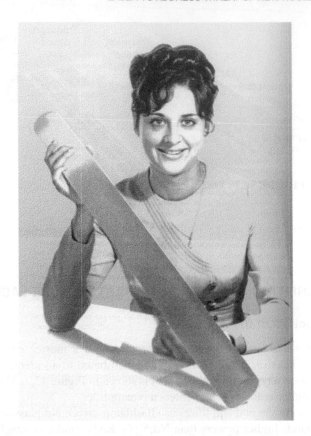

**FIGURE 13.2**　Nd:glass rod in 1970 French inertial confinement laser and amplifier.

Flash tubes　　　Disks

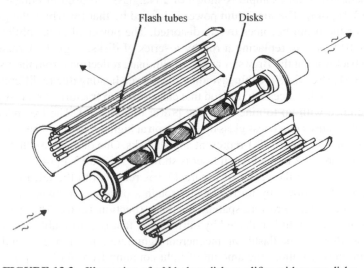

**FIGURE 13.3**　Illustration of a Nd:glass disk amplifier with many disks.

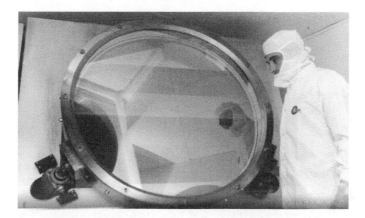

**FIGURE 13.4**   Nd:glass disk in 1974 French inertial confinement laser and amplifier.

flash light subsides. This means allowing the light to pass through the main amplifier four times before setting the polarization to switch the light out of the cavity through mirror LM3. A total of 7680 flashlamps are used in the 48 beamlines throughout the system, using a total electrical energy of 400 MJ from capacitor banks. As a result, even if nuclear fusion occurs, there is unlikely to be net energy produced.

The beam travels back to the right through the power amplifier and the central transport spatial filter. The spatial filters and deformable mirrors are critical to recovering the spatial coherence lost in distortion by the amplifiers. Figure 13.5 [127] shows 24 beamlines on one of the two parallel identical laser bays of the NIF facility; the people in the left foreground give a sense of the immensity of the laser.

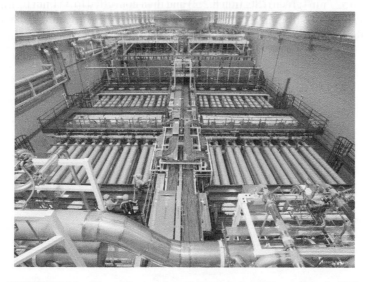

**FIGURE 13.5**   One of the two laser bays showing 24 beamlines in NIF.

**FIGURE 13.6**   Target of the beams during assembly.

Transport optics in Figure 13.1 now move the 48 beams to their positions around a 10 m diameter spherical target chamber to place more than a megajoule of energy on a single centimeter-size spot at the center of the sphere at precisely the same time within picoseconds ($10^{-12}$ s) (Figure 13.6) [127]. This explains the enormity of the difficulty involved.

In a final step (not shown in Figure 13.6) before merging beams, because ultraviolet penetrates the plasma better than infrared, the wavelength is halved from 1053 nm (infrared) to 527 nm (green) (Section 8.2.2) and then converted to 351 nm (ultraviolet) using thin sheets cut from a single crystal of potassium dihydrogen phosphate. The recovery time to cool down after firing the laser is expected to be around 5 hours.

# CHAPTER 14

# PROTECTING ASSETS FROM DIRECTED ENERGY LASERS

Military targets must be protected from high-energy light pulses that propagate efficiently across space at high speed and have potential to cause damage [103]. Coating targets can reduce the damage for some combinations of weak laser source and robust target but in general we need laser warning devices (LWD). A laser warning system was built into the system to protect aircraft from homing missiles in Section 12.3. The problem is somewhat similar to avoiding a speeding ticket in your car by using a laser warning detector, except perhaps that the military target may be worth hundreds of millions of dollars and many lives may be involved. The challenge is to detect the threat laser light fast enough for the target to evade the threat beam or launch countermeasures to save the target.

Almost every scenario is unique because targets include personnel, vehicles, missiles, aircraft, or satellites, while the threat could be from a variety of lasers: power lasers (Chapter 8), pulsed lasers (Chapter 9) and, ultrahigh-power lasers (Chapter 10). Sometimes the target is large relative to the threat laser beam diameter, a ship for example, and many coordinated sensors are required on the target. In ocean environments, the threat laser can scatter to form haze and fewer sensors may suffice. The laser threat beam may be modulated to provide unique identification for that beam.

*Military Laser Technology for Defense: Technology for Revolutionizing 21st Century Warfare*,
First Edition. By Alastair D. McAulay.
© 2011 John Wiley & Sons, Inc. Published 2011 by John Wiley & Sons, Inc.

Detecting laser threats in the future may become more difficult because new cyclotron-based lasers such as the free-electron laser (Chapter 10) [126] are tunable, orders of magnitude more powerful, and can be used at any microwave or light frequency.

In Section 14.1, we discuss the characteristics of lasers, some of which are measured by the laser warning device. In Section 14.2, we describe four basic approaches for laser warning devices from which systems may be tailored for specific applications.

## 14.1   LASER CHARACTERISTICS ESTIMATED BY LASER WARNING DEVICE

The main types of threat laser for military applications include laser diodes, diode-pumped solid-state lasers, fiber lasers, chemical lasers, and free-electron lasers (Section 8.1.3). A laser warning system is designed to detect one or more of the following relevant laser characteristics:

- *Frequency and Power of Light*: Fortunately, with the exception of the free-electron laser, there are a finite number of frequencies that can be easily generated with a laser because frequency is determined by the energy of the photon resulting when an electron falls across a bandgap from a higher energy level to a lower one in a lasing material. The number of low-cost easy-to-process lasing materials capable of generating high power is limited. Consequently, in detecting current high-power lasers we have only a few frequencies to consider.

  The application also influences the choice of laser frequency; for example, target designators where an operator points a beam at a target to relay its GPS coordinates require visible light. In contrast, an infrared (IR) range finder, not visible to the eye, is less likely to alert the target. In some cases, eye-safe lasers are desirable, similar to those operating around 1550 nm used in optical telecommunications. Minimizing the affects of the atmosphere can also influence frequency selection (Section 16.1).

- *Bandwidth or Temporal Coherence*: The bandwidth of the light, linewidth for a laser, indicates how close the light is to a single frequency. A narrow linewidth or equivalently a high temporal coherence (Section 6.1) defines the presence of a laser and may provide insight into the laser and its application. Communication lasers may have narrow linewidth so that many wavelength division multiplexed signals can pass through an optical amplifier. Detecting laser light is more difficult in high ambient light such as sunshine [26].

- *Direction of Laser and Spatial Coherence*: Spatial coherence is related to the rate at which a laser beam spreads with diffraction (Sections 3.2.2 and 1.3.6) and depends on the size of the emitting surface of the laser. Direction of the beam can be determined more precisely for a beam that is spreading more slowly. Beam direction systems are related to wave front detectors used in adaptive optics (Section 5.3.1). The direction of the source can indicate whether the laser is from an airborne vehicle such as a drone or airplane, from a satellite, or from a ground vehicle. Identifying the direction aids the threat assessment and permits immediate retaliation.

• *Pulse and Modulation*: Pulses are commonplace for laser weapons because for the same average laser power, greater damage may be inflicted by the higher peak power. A laser warning detector must respond faster to detect shorter pulses. Unfortunately, fast response conflicts with high sensitivity, which requires long integration times. The platform and application often dictate which is more critical, nevertheless both types of system may be required in the same installation. Many laser beams are modulated in the battlefield to avoid confusion with other beams, for example, target designators and for communications.

## 14.2 LASER WARNING DEVICES

In order to identify a laser threat and select an appropriate response, we simultaneously estimate several parameters for the light impinging on a laser weapon warning system. First, accurately estimating the direction of the source allows opportunity for an immediate response to neutralize the source. Second, estimating power and frequency enables us to assess the goal of the laser source and its damage potential from which we can decide on the threat level and corresponding evasion or retaliation response. Third, the bandwidth in frequency of the source $\Delta f$ indicates the level of temporal coherence $t_c - 1/\Delta f$ and can provide data on the modulation and pulse width. Indeed, laser light is distinguished from ambient light by its high temporal and spatial coherence [26]. The subject of spectrum analysis was discussed in several places in the book (Sections 4.2 and 15.3) and many of the techniques presented there can be used. The array waveguide structure in Section 14.2.4 provides a compact integrated optic spectrum analyzer.

We describe four approaches that can be used in different circumstances. The simultaneous estimation of direction and frequency with a grating (Section 14.2.1) is fast but has limited sensitivity to low light levels. Performance may be improved by separating the tasks of estimating direction and frequency. A single lens can replace the grating to provide direction only (Section 14.2.2), as in wave front detection in adaptive optics (Section 5.3.1). This is more sensitive to low light level because all the light is used purely for direction estimation. An interferometer (Chapter 6) [16] provides more accurate measurement of frequency only [28] but measures a single frequency unambiguously only within a spectral range. In Section 14.2.3, we select a Fizeau interferometer, because it can simultaneously estimate multiple frequencies unambiguously without mechanical motion. This allows bandwidth measurement and will operate on very short pulses, assuming fast detectors. However, now a curved mirror is needed to direct the beam into the interferometer. In Section 14.2.4, we describe an integrated optic spectrometer that is more robust, accurate, and less expensive. Such spectrometers will be required for new cyclotron-based lasers such as the free-electron laser that can operate at any frequency and are typically tunable over approximately 20% range.

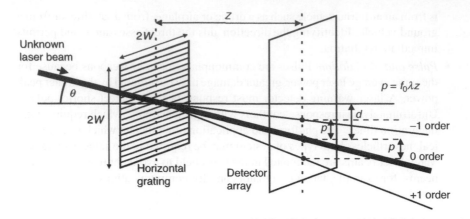

**FIGURE 14.1**   Intensity field from a sinusoidal grating in a laser warning device.

We note that laser beams have a cross-sectional Gaussian intensity distribution that must be included for accurate analysis. Furthermore, beams that carry energy over kilometers through the atmosphere often have their diameter expanded to tens of centimeters to reduce the effects of atmospheric turbulence [88, 90]. For Gaussian beams and statistical detection and estimation techniques, see Section 2.1 and Chapter 5.

### 14.2.1   Grating for Simultaneously Estimating Direction and Frequency

A grating is a fast relatively inexpensive means of detecting and simultaneously estimating direction and frequency for a laser beam [177]. Figure 14.1 shows a ray of the plane wave front from laser light striking a transmissive cosinusoidal grating of size $2W \times 2W$ at an angle $\theta$ to the grating normal in a laser warning device. Light is diffracted by the grating and propagates to a detector array.

As in Section 4.2.2, equation (4.2), the transmission function in the $z$ direction for the square cosinusoidal grating [49, 83] is (Figure 14.2).

$$U_{\text{in}}(x, y) = \left[ \frac{1}{2} + \frac{1}{2} \cos(2\pi f_0 x_1) \right] \text{rect} \left( \frac{x_0}{2W} \right) \text{rect} \left( \frac{y_0}{2W} \right) \qquad (14.1)$$

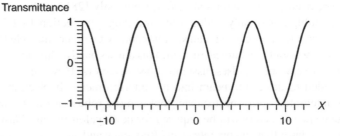

**FIGURE 14.2**   Cosinusoidal grating shape.

The square bracket represents the sinusoidal grating pattern of spatial frequency $f_0$ in the $x$ direction. The first $\frac{1}{2}$ in equation (14.1) provides the mean or zero-order diffraction (straight through) in Figure 14.1, which arises because intensity cannot be negative. The rectangular functions represent the finite $2W \times 2W$ square aperture of the grating.

Incident light with propagation constant $k$ at an angle $\theta$ with the horizontal has a downward propagating component at the grating of $\exp\{jk \sin\theta\, x_1\} = \exp\{j2\pi x_1 \sin\theta/\lambda\}$. The normal incidence case, $\theta = 0$, was analyzed in Section 4.2.2. So in a similar manner to Section 4.2.2, the Fourier transform in the $x$–$y$ plane of the intensity immediately after the grating is, from equation (14.1),

$$U_{\text{out}}(x_0, y_0) = \mathcal{F}\left[U_{\text{in}}(x_1, y_1)\exp\left\{\frac{j2\pi x_1 \sin\theta}{\lambda}\right\}\right]$$

$$= \mathcal{F}\left[\frac{1}{2} + \frac{1}{2}\cos 2\pi f_0 x_1\right] * \mathcal{F}\left[\text{rect}\left(\frac{x_0}{2W}\right)\text{rect}\left(\frac{x_0}{2W}\right)\right] * \mathcal{F}\left[\exp\left\{\frac{j2\pi \sin\theta\, x_1}{\lambda}\right\}\right]$$

$$= \left[\frac{1}{2}\delta(f_x f_y) + \frac{1}{4}\delta(f_x + f_0, f_y) + \frac{1}{4}\delta(f_x - f_0, f_y)\right]$$

$$* (2W)^2 \text{sinc}(2Wf_x)\text{sinc}(2Wf_y) * \delta\left(f_x + \frac{\sin\theta}{\lambda}\right)$$

$$- \left(\frac{2W}{2}\right)^2 \text{sinc}(2Wf_y)\left[\text{sinc}\left\{2W\left(f_x + \frac{\sin\theta}{\lambda}\right)\right\}\right.$$

$$\left. + \frac{1}{2}\text{sinc}\left\{2W\left(f_x + \frac{\sin\theta}{\lambda} + f_0\right)\right\} + \frac{1}{2}\text{sinc}\left\{2W\left(f_x + \frac{\sin\theta}{\lambda} - f_0\right)\right\}\right] \quad (14.2)$$

If a front-end telescope is used or the beam from the laser is less than the grating aperture size, the Gaussian nature of the laser beam should be included as discussed in Section 14.2.2.

The three orders in the square bracket in the last line of equation (14.2) correspond to those in Figure 14.1 (0, −1, 1). Assuming sufficient spacing to avoid overlap of diffraction orders, we can write the Fraunhofer diffraction far-field intensity at the output by introducing scaling for optics, $f_x = x_0/(\lambda z)$ and $f_y = y_0/(\lambda z)$, and then multiplying by its conjugate,

$$I_{\text{out}}(x_0 y_0) = \frac{1}{(\lambda z)^2}\left\{\frac{1}{2}(2W)^2\right\}^2 \text{sinc}^2\left(\frac{2Wy_0}{\lambda z}\right)$$

$$\left[\text{sinc}^2\left(\frac{2W}{\lambda z}(x_0 + z\sin\theta)\right) + \frac{1}{4}\text{sinc}^2\left\{\frac{2W}{\lambda z}(x_0 + z\sin\theta + f_0\lambda z)\right\}\right.$$

$$\left. + \frac{1}{4}\text{sinc}^2\left\{\frac{2W}{\lambda z}(x_0 + z\sin\theta - f_0\lambda z)\right\}\right] \quad (14.3)$$

The intensity of the far field from a cosinusoidal grating, equation (14.3), at a far-field plane, is illustrated in Figure 14.1. Comparison of equation (14.3) with the normal incidence case, equation (4.5), shows that the output orders are shifted down by $d$ in

Figure 14.1 compared to the normal incident case (Figure 4.4). Information about the wavelength λ of the unknown laser is obtained from the distance $p$ on the detector array in Figure 14.1.

$$\lambda = \frac{p}{f_0 z} \tag{14.4}$$

The distance between grating and detector array, $z$, is known for the system. Information about the unknown laser direction $\theta$ is obtained from the distance $d$ in the output detector array:

$$\tan \theta = \frac{d}{z} \tag{14.5}$$

Due to atmospheric turbulence, the image at the detector will wander around and the focus regions fluctuate in size with time. In this case, the equation approximates the centroid of the pattern on the image plane. The turbulence effect is in addition to that caused by relative motion of the target and beam source.

### 14.2.2 Lens for Estimating Direction Only

A lens can be used in place of a grating to avoid the inefficiencies of the sinusoidal grating due to dispersion into three orders. Hence, the lens direction finding system is more sensitive than a grating for finding the direction of low light level laser sources such as those arising in munitions guiding lasers or due to scattering from the atmosphere. This is similar to a Hartmann wave front sensing in adaptive optics (Section 5.3.1) [144, 163]. Figure 14.3 shows a Gaussian beam of laser light with spot size $W_{lens}$ striking a lens at an angle $\theta$ with the horizontal in a laser warning device.

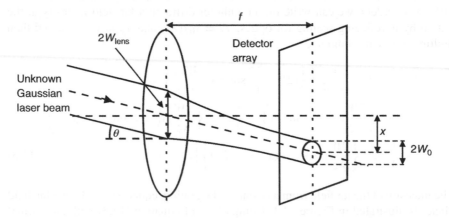

**FIGURE 14.3**   A lens-based direction finding system in a laser warning device.

A Gaussian beam is fully defined by its spot size $W(z)$ (radius of $1/e$ amplitude spot) and the radius of curvature of the phase front $R(z)$ (Section 2.1.1, equation (2.18)) times $\exp\{-jkz\}$, equivalent to Equation (2.19).

$$U(x, y, z) = \frac{A_1}{\omega} \frac{W_0}{W(z)} \exp\left\{-j[kz - \zeta(z)] - \rho^2\left[\frac{1}{W(z)^2} + \frac{jk}{2R(z)}\right]\right\} \quad (14.6)$$

where $k = 2\pi/\lambda$ is the propagation phase constant, $W_0$ is the beam spot size at the waist (narrowest point), $\zeta$ is the beam divergence angle, and $\rho$ is the radial direction in the transverse plane. In Figure 14.3, a lens is used to focus a Gaussian beam of spot size $W_{lens}$ to its narrowest spot size (waist) $W_0$ at the detector (Figure 14.3) [148, 176].

On passing a Gaussian beam through a lens of focal length $f$, the spot size $W_{lens}$ remains the same but the slope is reduced by $1/f$, equation (1.22). As slope equals the reciprocal of radius of curvature $R$, the radius of curvature of the Gaussian beam exiting the convex lens is given by

$$\frac{1}{R} = \frac{1}{R_{in}} - \frac{1}{f} \quad (14.7)$$

Assuming the incoming beam approximates a plane wave because of its great distance from the source, $R_{in} = \infty$, then from equation (14.7) the radius of curvature of the Gaussian beam leaving the lens is $R = -f$: the negative sign represents a converging beam. The Gaussian beam parameters for the beam exiting the lens are $W = W_{lens}$ and $R = -f$. The Gaussian beam parameters at the detector array are radius of curvature $R = \infty$ and the waist size $W_0$ that can be derived from the equations in Section 2.1.3 item (d). Substituting for $z$, from equation (2.34) into equation (2.43) and using the binomial approximation gives the beam size $W_0$ on the detector.

In the same manner as for the grating (Section 14.2.1), the image on the detector array moves around and the spot size varies because of atmospheric turbulence. Consequently, the centroid of the spot must be computed to determine the direction of the laser source.

## 14.2.3   Fizeau Interferometer

A Fizeau interferometer [28] has two off-parallel planar reflecting surfaces separated by a small angle $\phi$ (Figure 14.4). At any point along $z$, the plates approximate a parallel plate because $\phi$ is small. So we first consider the parallel plate interferometer in Figure 14.5 in which an unknown laser beam in air impinges at an angle $\alpha$ and is reflected from the top and bottom surfaces. We can determine the increased path length $\Delta s$ of the bottom surface reflection relative to the top surface reflection in two parts: the part in the glass $d_g$ and the part $d_p$ required to bring the phases together for

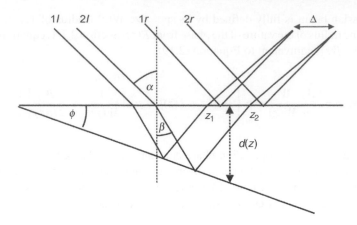

**FIGURE 14.4**    Fizeau interferometer.

the output beam.

$$\Delta s = d_g - d_p = \frac{2nd}{\cos \beta} - 2d \tan \beta \, \sin \alpha = \frac{2nd}{\cos \beta} - 2d \tan \beta \, n \, \sin \beta = 2nd \, \cos \beta$$

$$(14.8)$$

where we used Snell's law $\sin \alpha = n \sin \beta$ and $1 - \sin^2 \theta = \cos^2 \theta$. When the difference in path lengths $\Delta s = m\lambda$ ($m$ an integer), light from the two paths is in phase and combines constructively for a maximum intensity, while when the difference in path lengths is $\Delta s = \lambda/2 + m\lambda$, light from the two paths is out of phase and combines destructively for a minimum intensity. Setting $\Delta s = m\lambda$ in equation (14.8) gives

$$m\lambda = 2nd \, \cos \beta \qquad (14.9)$$

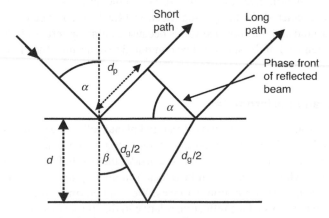

**FIGURE 14.5**    Parallel plate interferometer.

In the Fizeau interferometer, the spacing between the surfaces $d(z)$ changes with distance $z$ along the wedge formed and this changes the resonant wavelength $\lambda$ with distance. By detecting wavelengths of interest in parallel, the Fizeau is more suitable for rapid detection of pulsed laser sources than a resonator that detects only a single frequency. From equation (14.9) we can write for adjacent peaks,

$$2n[d(z_2) - d(z_1)]\cos \beta = \lambda \quad \text{or} \quad d(z_2) - d(z_1) = \frac{\lambda}{2n \cos \beta} \qquad (14.10)$$

The distance $\Delta$ along $z$ between two fringe maxima may be written from Figure 14.4 and equation (14.10):

$$\Delta = z_2 - z_1 = \frac{d(z_2) - d(z_1)}{\tan \phi} = \frac{\lambda}{2n \cos \beta \tan \phi} \qquad (14.11)$$

Equation (14.11) shows that $\Delta$ varies with wavelength $\lambda$. Further as $\phi$ is small, the resonant wavelength continues to behave locally at location $z$ like a parallel plate for which $\lambda$ depends on plate separation $d(z)$. Consequently, changing $\lambda$ will both move the fringes and change the separation of their maxima.

For parallel plate resonators, $\lambda$ can be determined within only one spectral range because of higher order integers $m$ in equation (14.10) give the same output. The free spectral range, the frequency range between two maxima, is $\delta v = c/\Delta s$. Inserting for $\delta v$ from the second equation (6.20) that converts from frequency range $\delta v$ to wavelength range $\delta \lambda$ gives

$$\frac{c}{\lambda^2}\delta \lambda = c/\Delta s \qquad (14.12)$$

or, replacing $\Delta s$ using equation (14.8),

$$\delta \lambda = \frac{\lambda^2}{\Delta s} = \frac{\lambda^2}{2nd \cos \beta} \qquad (14.13)$$

Compared to a parallel plate interferometer, the Fizeau interferometer is better for handling pulsed lasers because it measures many wavelengths simultaneously as a function of distance $z$ along the interferometer. Hence we can also estimate absolute wavelength over a wide range of wavelengths and avoid the ambiguity inherent in the periodicity of the parallel plate interferometers [28].

## 14.2.4 Integrated Array Waveguide Grating Optic Chip for Spectrum Analysis

The laser threat detector must operate with current laser weapon designs as well as future cyclotron-based lasers that we expect will initially use the same wavelengths as current laser weapons because of availability of components and instrumentation. The wavelengths of current and anticipated laser weapons are listed in Table 8.1. For laser threat detection, we propose a robust inexpensive integrated optic spectrometer

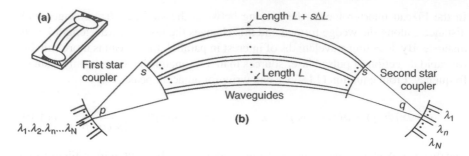

**FIGURE 14.6**   Photonic integrated circuit layout for array waveguide grating.

to determine the laser wavelength [103]. The array waveguide grating (AWG) is an integrated optic version of the grating (Section 14.2.1) as a spectrometer for estimating wavelength. The spectrometer consists of an array waveguide grating similar to those used in telecommunications (Figure 14.6) [2, 53, 75].

The array waveguide grating is used for wavelength division multiplexing in telecommunications. Prior to the invention of the AWG, a separate filter was used for each wavelength, typically 128 per band. The AWG handles all 128 different frequency channels at once. We propose an AWG as a spectrum analyzer for detecting laser threats such as those from new tunable high-power free-electron lasers (Section 11.1) [103] or other cyclotron-based lasers (Chapter 10).

Figure 14.6a shows an integrated optic AWG chip and Figure 14.6b shows the circuit layout for the AWG chip. As a spectrometer, the input with multiple wavelengths is entered at one of the input ports at lower left, for example, the $p^{th}$ input port. The wavelengths are spread to separate ports at the output at the right-side. There are three parts: a first star coupler, a waveguide part, and a second star coupler.

In the first star coupler, the input enters the coupler and spreads out due to diffraction; thus, the input is broadcast to all the outputs of the first star coupler. The circular curvature of the first star coupler ensures that all the outputs from the first star coupler enter the waveguides in the waveguide section with the same phase. The waveguide section has waveguides of increasing lengths. The difference in length between a waveguide and its neighbor is $\Delta L$. In the second star coupler, the light at wavelength $\lambda_1$ is focused to the first output waveguide, $\lambda_2$ to the second, and so on.

The operation can be understood by examining phase delays throughout the system. As the upper waveguides are longer than those below, light is delayed in the upper guides relative to the lower ones. The amount of delay depends on wavelength. Therefore, at the entrance to the second star coupler, the wave front defined by the phase peaks has a tilt depending on wavelength. Similar to steering with a phased array antenna, the tilt causes beams of different wavelengths to focus to different outputs. We derive two design equations for the AWG: the first computes the relationship between the incremental length $\Delta L$ and the nominal wavelength $\lambda_0$, and the second computes the relationship between the coupler geometry and the resolution between wavelengths at the output $\Delta\lambda$.

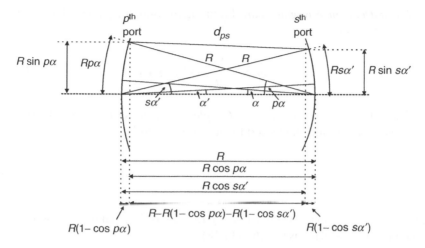

**FIGURE 14.7**   First star coupler geometry.

Figure 14.7 defines the star coupler geometry. The circular boundaries in the star coupler have radius $R$ and the distance separating the two arcs on the axis is also $R$. The angles subtended by the first ports above the axis at the opposite faces are $\alpha$ and $\alpha'$ for the two directions, respectively, as shown. The angle subtended at the output axis by the arc to the $p^{\text{th}}$ input port is $p\alpha$ and that subtended by the $s^{\text{th}}$ output port at the input axis is $s\alpha'$. In order to compute the phase lag from the $p^{\text{th}}$ input port to the $s^{\text{th}}$ output port, we first compute the distance $d_{ps}$ using Pythagoras theorem with horizontal and vertical distances. The horizontal distance is computed by dropping a perpendicular from the input $p^{\text{th}}$ port and output $s^{\text{th}}$ port to the axis. The horizontal distance between the ports is the distance between these lines $R[1 \quad (1 - \cos(p\alpha)) - (1 - \cos(s\alpha'))]$. We drop perpendiculars from the $p^{\text{th}}$ port and $s^{\text{th}}$ port to the vertical line through the intersections of curved surface and axis to obtain the vertical distance $R(\sin(p\alpha) - R\sin(s\alpha'))$. Hence, by Pythagoras theorem

$$d_{ps}^2 = R^2(-1 + \cos(p\alpha) + \cos(s\alpha'))^2 + R^2(\sin(p\alpha) - \sin(s\alpha'))^2 \qquad (14.14)$$

For small angles $p\alpha$ and $s\alpha'$, $\cos(p\alpha) \to 1$, $\cos(s\alpha') \to 1$, $\sin(p\alpha) \to p\alpha$, and $\sin(s\alpha') \to s\alpha'$ and neglecting second-order terms $(p\alpha)^2$ and $(s\alpha')^2$, and using the binomial theorem to remove the square root, equation (14.14) becomes

$$d_{ps} = R(1 - ps\alpha\alpha') \qquad (14.15)$$

For the three parts of the AWG, we write the following:

1. *Phase Delay Across First Star Coupler.* From equation (14.15), using phase $\phi = kd = 2\pi n_{\text{coupler}}/\lambda$, the phase delay from input port $p$ to output ports is,

$$\phi_{ps} = \frac{2\pi}{\lambda/n_{\text{coupler}}} R(1 - ps\alpha\alpha') \qquad (14.16)$$

2. *Phase Delay in Waveguide Part.* From Figure 14.6, the phase delay along the $s^{\text{th}}$ waveguide is

$$\phi_s = \frac{2\pi}{\lambda/n_{\text{wg}}}(s\Delta L + L) \qquad (14.17)$$

3. *Phase Delay Across Second Star Coupler.* Analogous to equation (14.16), the phase delay for the input $s$ port to the output $q$ port is

$$\phi_{sq} = \frac{2\pi}{\lambda/n_{\text{coupler}}} R(1 - sq\alpha\alpha') \qquad (14.18)$$

The total phase from the AWG input port $p$ to the AWG output port $q$ is written by adding phase from equations (14.16)–(14.18):

$$
\begin{aligned}
\phi_{p,s,q} &= \phi_{ps} + \phi_s + \phi_{sq} \\
&= \frac{4\pi}{\lambda/n_{\text{coupler}}} R + \frac{2\pi}{\lambda/n_{\text{wg}}}(s\Delta L + L) \\
&\quad - \frac{2\pi}{\lambda/n_{\text{coupler}}} R\alpha\alpha' s(p + q)
\end{aligned}
\qquad (14.19)
$$

where the first term of the last equation sums the first and constant terms from equations (14.16) and (14.18) and the last term sums the second terms of these equations for which there is port number dependence.

The phase difference between two waveguide paths, $s$ and $s - 1$, from equation (14.19) removes constant terms:

$$
\begin{aligned}
\Delta\phi_{pq} &= \phi_{p,s,q} - \phi_{p,s-1,q} \\
&= \frac{2\pi}{\lambda/n_{\text{wg}}} \Delta L - \frac{2\pi}{\lambda/n_{\text{coupler}}} R\alpha\alpha'(p + q)
\end{aligned}
\qquad (14.20)
$$

The power at the output port $q$ results from summing signals with phase $\Delta\phi_{pq}$ for all $M$ waveguides ($s$ from 0 to $M - 1$), (see Section 9.2.1)

$$
\begin{aligned}
P_{pq} &= \frac{P_{\text{in}}}{M^2} \left| \sum_{s=0}^{s=M-1} \exp\{js\Delta_{pq}\} \right|^2 \\
&= \frac{P_{\text{in}}}{M^2} \left| \frac{1 - \exp\{jM\Delta\phi_{pq}\}}{1 - \exp\{j\Delta\phi_{pq}\}} \right|^2 \\
&= \frac{P_{\text{in}}}{M^2} \frac{\sin^2(M\Delta\phi_{pq}/2)}{\sin^2(\Delta\phi_{pq}/2)}
\end{aligned}
\qquad (14.21)
$$

When $\Delta\phi_{pq} \to 2\pi$, all paths are in phase at the output, the angles are small and $\sin\theta \to \theta$, so the maximum power approaches

$$P_{pq}(\Delta\phi_{pq}) \to \frac{P_{in}}{M^2} \frac{(M\Delta\phi_{pq}/2)^2}{(\Delta\phi_{pq}/2)^2} = P_{in} \tag{14.22}$$

Substituting $\Delta\phi_{pq} = 2\pi$ into the left-hand side of equation (14.20), canceling $2\pi$, and setting $\lambda = \lambda_{pq}$,

$$\lambda_{pq} = n_{wg}\Delta L - n_{coupler} R\alpha\alpha'(p+q) \tag{14.23}$$

$\lambda_{pq}$ is the wavelength for which all paths through different waveguides combine at the output port $q$. We can write equation (14.23) by separating out and defining a constant term $\lambda_0$ from a term $\Delta\lambda$ that depends on the input port $p$ and output port $q$:

$$\lambda_{pq} = \lambda_0 - (p+q)\Delta\lambda \tag{14.24}$$

where the newly defined terms provide the two design equations for an AWG. The first design equation for an AWG relates the operating wavelength $\lambda_0$ to the incremental length $\Delta L$ of the waveguides:

$$\lambda_0 = n_{wg}\Delta L \tag{14.25}$$

and the second design equation for an AWG relates the wavelength resolution, spacing between wavelengths, $\Delta\lambda$, and the geometric values for the layout, $\alpha$ and $\alpha'$:

$$\Delta\lambda = n_{coupler} R\alpha\alpha' \tag{14.26}$$

### 14.2.5  Design of AWG for Laser Weapons

Assuming refractive index of both star coupler and waveguides is $n$, from equation (14.25), we write the first design equation, the difference in length between two adjacent central waveguides, as

$$\Delta L = \frac{\lambda_0}{n} \tag{14.27}$$

From equation (14.26) for the second design equation, we assume that the angle $\alpha = \alpha'$ for a given device of radius $R$ and spacing between wavelengths $\Delta\lambda$:

$$\alpha = \sqrt{\frac{\Delta\lambda}{nR}} \tag{14.28}$$

For illustration and to show feasibility with modern lithography machines, we design a device with a center wavelength at a laser weapon wavelength around

$\lambda_0 = 1.5\,\mu$m and a range of $\pm 15\%$, comparable to that of a cyclotron-based laser, such as a free-electron laser. The $\pm 15\%$ is an estimate for the tunable range of a specific free-electron or other cyclotron laser. For reference, the airborne laser uses a wavelength of 1.315 $\mu$m. For the laser weapon at a wavelength $\lambda_0 = 1.5\,\mu$m, from equation (14.27) the distance in length between adjacent waveguides is $\Delta L = 1.5 \times 10^{-6}/1.5 = 10^{-6}$ or 1 $\mu$m for waveguide refractive index $n = 1.5$. Current lithographic machines for telecommunication AWGs have the capability to provide 1 $\mu$m resolution in waveguide length.

For $\pm 15\%$ tuning around $\lambda_0 = 1.5\,\mu$m, the wavelength range of operation is from 1.275 to 1.725 $\mu$m or 450 nm. If 200 wavelength bin slots are selected for the array waveguide spectrometer, each slot is $\Delta\lambda = 450/200 = 2.25$ nm wide. For a radius of $R = 5$ cm, using equation (14.28), the angle $\alpha$ in Figure 14.7 is $\alpha = \sqrt{2.25 \times 10^{-9}/(1.5 \times 5 \times 10^{-2})} = 1.7 \times 10^{-4}$ or 0.17 mrad. This is a reasonable angle for current lithography because it is larger than that used in current telecommunication array waveguides [2, 75].

# CHAPTER 15

# LIDAR PROTECTS FROM CHEMICAL/ BIOLOGICAL WEAPONS

Lidar stands for light detection and ranging and refers to systems in which a beam of laser light shines into the atmosphere and is scattered or reflected from an object or from a cloud [168]. Light wavelengths are short enough to interact with chemical and biological elements in the air, so the backscatter from an aerosol cloud may be analyzed for the presence of chemical and biological weapons. This chapter focuses on lidar for detecting the presence of chemical and biological weapons in the atmosphere.

Other military applications involve scanning a pulsed laser beam to produce 3D images of a target. The depth is obtained by measuring the time between the emission of a pulse and receiving the backscattered pulse. When scanning the earth from an aircraft or a satellite, 3D remote sensing surveillance maps are obtained.

In Section 15.1, we introduce lidar for detecting and assessing threats from chemical and biological weapons and briefly mention other military applications. In Section 15.2, we describe a typical lidar system. In Section 15.3, we describe optical spectrum analyzers for identifying chemicals from lidar return signals. In Section 15.4, we consider a lidar system with spectrum analyzer to determine the presence of chemical or biological weapons and the associated threat assessment.

*Military Laser Technology for Defense: Technology for Revolutionizing 21st Century Warfare*, First Edition. By Alastair D. McAulay.
© 2011 John Wiley & Sons, Inc. Published 2011 by John Wiley & Sons, Inc.

## 15.1   INTRODUCTION TO LIDAR AND MILITARY APPLICATIONS

Lidar is developing fast to become one of the more important technologies for military applications. Lidar, which operates at light frequencies, is similar in concept to radar (radio detection and ranging) that operates at radio frequencies or microwaves. Lidar, in addition to 3D image scanning capabilities, has a significant additional capability compared to radar: it can identify chemicals. The shorter light wavelengths that range from infrared through visible to ultraviolet can match the size of chemical or biological molecules and particles in an aerosol. Both absorption and scattering depend on particle size and concentration. For absorption, when a match occurs the molecule or particle resonates, absorbing energy and causing a dip in intensity in the backscattered signal at the resonant frequency for that molecule or particle. Sweeping the light through a range of wavelengths allows spectral identification and concentration of the chemicals present in the air or in a cloud. This chapter focuses on the ability of lidar to remotely measure chemicals in the atmosphere.

Lidar can profile the atmosphere up to over 100 km from the ground and higher from aircraft or satellites. Lidar atmospheric measurements include wind, temperature, humidity, trace gases, clouds, and aerosols. These parameters can drastically affect performance of military systems including missiles, artillery, lasers, communications, imaging, visible range, and aircraft motion. Lidar is used to estimate turbulence for adaptive optics (Section 5.3.1.4) and for optimizing windmill location and blade direction. Ladar can detect the presence of an aerosol cloud in the sky.

Once an aerosol is detected, lidar can be used to estimate the extent, motion, concentration, and composition of an aerosol cloud: an aerosol is a system of colloidal particles in a gas, smoke, or fog. Colloidal particles have a diameter of $10^{-7}$ to $10^{-5}$ cm, larger than that of most inorganic molecules, and remain suspended indefinitely, due to the motion and associated electrostatic charges in the surrounding media. Detection of trace gases, toxic chemicals, and biological agents and their concentrations from the lidar enables a military threat assessment. The responses must then be evaluated: for example, is contamination equipment needed? Can we avoid the cloud? Is the aerosol hiding something? Do we need to use radar such as W-band radar or 94 GHz radar in Chapter 16 to see into it? Can we operate through it? A recent startling example of our unpreparedness for unpredicted aerosol events is illustrated by the dilemma regarding whether jet aircraft can safely fly through airborne volcanic ash and where and how long the ash will interfere with international travel. The molten iron, the crust of which we live on, is in turmoil, indicated by the recent beginning of reversal of the earth's magnetic field; we can expect a great deal more problems of penetration of the crust leading to earthquakes and volcanoes.

### 15.1.1   Other Military Applications for Lidar

Lidar can be used like radar but the shorter relative wavelength on a visually clear day (Section 16.1) provides narrower beams, higher space and time resolution, and, by timing pulse response, the ability to perform 3D profiling of targets for identification

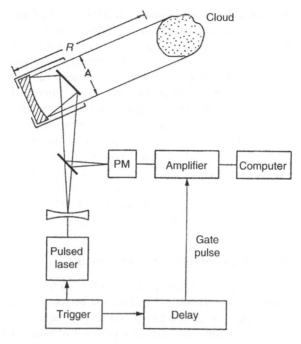

**FIGURE 15.1**   Lidar system.

[142]. In this context and in military systems, the word ladar is often used in place of the word lidar: ladar stands for laser detection and ranging. A scanning lidar is used in robotics to estimate distance to objects to aid image segmentation in image processing. Scanning lidar is also used for remote sensing and profiling of the earth from aircraft or satellites [168].

## 15.2   DESCRIPTION OF TYPICAL LIDAR SYSTEM

Figure 15.1 [28] shows a block diagram of a typical lidar system for investigating chemical and biological constituents in an aerosol cloud. A triggered pulsed laser shines light toward a cloud via a Nasmyth telescope (Section 1.3.4.2). The Nasmyth telescope, in which light is reflected to one side, is a variation of a Cassegrain telescope (Section 1.3.4.1), in which light passes through a hole in the primary mirror. The return pulse from the cloud is detected with a photomultiplier tube PM and gated in an amplifier by the trigger function delayed by the range for which response is sought. An optical spectrum analyzer and filters may precede the photomultiplier tube.

### 15.2.1   Laser

The laser is selected to match the application; in particular, the wavelength is criti-cal (Chapter 8) [166]. Currently, the most commonly used lasers are, for ultraviolet, excimer and, for infrared, Nd:YAG (Section 8.2) at 1064 nm. Nd:YAG can be doubled to 532 nm (Section 8.2.2), tripled to 355 nm, and quadrupled to 266 nm. Longer wavelengths are obtained by stimulated Raman scattering in hydrogen and deuterium that when applied to an excimer laser provides ozone differential absorption lidar and solar-blind lidar. In solar-blind lidar, superior performance is achieved in bright sunshine by using ultraviolet wavelengths not emitted by the sun and filtering the sunlight out. For sweeping through wavelengths for chemical analysis with differential absorption and resonance fluorescence lidars, dye lasers pumped with excimer or Nd:YAG are being replaced with tunable solid-state lasers such as titanium:sapphire with optical parametric amplifiers [176]. Infrared lidars suitable for Doppler lidar involve doping of crystalline lattices YAG, YLF or LuAG with active dopants such as Nd, Ho, Tm, Cr, Er, or Yb. New lasers are routinely considered for lidar as they are developed; for example, fiber lasers emitting eye-safe 1500 nm are valuable for some high-power applications and free-electron lasers allow tuning over limited ranges (Section 11.1).

### 15.2.2 Cassegrain Transmit/Receive Antennas

A Cassegrain or Nasmyth mirror telescope up to a few meters in diameter is typically used to launch the lidar beam and receive the backscatter (Section 1.3.4.1, Figure 1.11). For chemical/biological lidar, the sensitivity required is such that separate antennas in a biaxial system are normally used for transmit and receive. The beam expansion (Section 1.3.2) of the inverted telescope reduces the divergence due to diffraction to 100 $\mu$ rad. A field stop (Section 1.3.4) in the receiver telescope focal plane reduces the field of view to a few hundred $\mu$ rad. The narrow transmit and receive beams reduce the interfering background light and the effects of multiple scattering. The small field of view improves selectivity in spectrum analyzers. For a ground lidar investigating far distance targets, the upper atmosphere, such as a radar for mapping debris and objects in satellite space (Section 16.3.2), a chopper, timed by the emit pulse repetition rate, can eliminate interference from scatterers at close distances.

In a biaxial system, the launch beam expander (inverse telescope) and the receiving telescope are separated by half the antenna width or more to reduce inadvertent coupling of powerful transmit signals directly into the receiver system. In this case, there is a receiver field-of-view overlap function $O(R)$ that may require a few kilometer range to the target area before overlap reaches 100%, depending on field of view and separation.

### 15.2.3 Receiver Optics and Detector

An optical filter at the passband of the laser can filter out interference from unwanted wavelengths. Other filters, such as polarization filters, can be used in conjunction with a spectrum analyzer (Section 15.3). A photomultiplier tube or avalanche photodiode converts light to an electronic signal. Both photomultiplier tube and less sensitive avalanche photodiode have sufficient sensitivity to provide Geiger counter operation for which a count of individual photons measures very weak backscatter light intensities.

## 15.2.4  Lidar Equation

We derive the lidar equation for the received lidar signal [168]. The lidar equation can be used for many applications but the backscatter and absorption equations must be modified for the different applications, remote chemical sensing, target mapping, and ground mapping.

### 15.2.4.1  *Range Bin*

The lidar sends a pulse of duration $\tau$ into the atmosphere. If $P_0$ is the average power in the pulse, the energy in a pulse is $E = P_0\tau$ and the illuminated length of a pulse in space is $c\tau$, where $c$ is the velocity of light. For a pulse repetition rate of $f_{rep}$, the average power emitted is $P = Ef_{rep}$.

Because the pulse has duration, each point in the received time trace shows the response from a range bin $\Delta R$ that we now compute. At a time $t$ after sending the pulse, we observe the return signal. The front edge of the pulse returns from a scatterer at a distance from the lidar of

$$R_1 = \frac{ct}{2} \tag{15.1}$$

where the factor 2 accounts for the two-way travel distance $2R_1$. The rear edge of the pulse leaves the emitter at $\tau$ after the front edge and returns from a distance

$$R_2 = \frac{c(t - \tau)}{2} \tag{15.2}$$

The range bin $\Delta R$ from which return signals are obtained for pulse duration $\tau$ is then

$$\Delta R = R_1 - R_2 = \frac{c\tau}{2} \tag{15.3}$$

The system performance characteristic $K$ is written as

$$K = P_0 \frac{c\tau}{2} A\eta \tag{15.4}$$

where the strength of the return signal is proportional to the area of the telescope $A = \pi(D/2)^2$ with $D$ the diameter of the telescope main mirror, $\eta$ is the overall system efficiency, $P_0$ is the average power in the pulse, and the range bin $\Delta R = c\tau/2$.

### 15.2.4.2  *Antenna Gain and $R^2$ Fall Off with Distance*

For isotropic scatterers, the radiated scattered intensity $I_s$ lies on the surface of a sphere of radius $R$ according to $4\pi R^2$. The collected light $I_c$ for an antenna of area $A$ as a ratio of scattered light is

$$\frac{I_c}{I_s} = \frac{A}{4\pi R^2} \tag{15.5}$$

The collected intensity falls off with distance as $R^2$ and $A/R^2$ is known as the perception angle for the light scattered at distance $R$. For bistatic lidar, we combine the overlap function $O(R)$, defined in Section 15.2.2, and the $R^2$ fall off into

$$G(R) = \frac{O(R)}{R^2} \tag{15.6}$$

### 15.2.4.3 Backscatter Coefficient

The backscatter coefficient, $\beta(R, \lambda)$, at distance $R$ and wavelength $\lambda$, represents the ability of the atmosphere to scatter light back into the direction from which it came ($\pi$ radians from the forward direction). For the $j$th particle in a collection of varying particles, with concentration of $N_j$ particles per unit volume, and particle cross section per unit solid angle $[d\sigma_{j,\text{scat}}/d(\Omega)](\pi, \lambda)$, the backscatter coefficient is written as

$$\beta(R, \lambda) = \sum_j N_j \left[ \frac{d\sigma_{j,\text{scat}}}{d\Omega} \right] (\pi, \lambda) \tag{15.7}$$

If the $N$ scatterers per unit volume are identical and isotropic (radiating over solid angle $d\Omega \rightarrow 4\pi$) with cross section $\sigma_{\text{scat}}$, from equation (15.7)

$$4\pi\beta = N\sigma_{\text{scat}} \tag{15.8}$$

If the cross-sectional area of the lidar beam at the scattering volume is $A_L$, from equation (15.3), the scattering volume is $V = A_L \Delta R = A_L c\tau/2$. From equation (15.8), the intensity of scattered light is proportional to

$$A_s = N\sigma_{\text{scat}} V = N\sigma_{\text{scat}} A_L \frac{c\tau}{2} \tag{15.9}$$

Using $A_s$ from equation (15.9) and $N\sigma_{\text{scat}}$ from (15.8), the ratio of intensity of scattered light $I_s$ to emitted light $I_0$ is

$$\frac{I_s}{I_0} = \frac{A_s}{A_L} = N\sigma_{\text{scat}} \frac{c\tau}{2} = \frac{4\pi\beta c\tau}{2} \tag{15.10}$$

Substituting $I_s$ from equation (15.5) into equation (15.10) gives the ratio of collected light in the receiver $I_c$ to the emitted light $I_0$ by the transmitter:

$$\frac{I_c}{I_0} = \frac{I_s}{I_0} \frac{I_c}{I_s} = \left( \frac{\beta c\tau}{2} \right) \left( \frac{A}{R^2} \right) = \left( \frac{A\beta c\tau}{2R^2} \right) \tag{15.11}$$

in terms of the perception angle $A/R^2$, backscatter coefficient $\beta$, and pulse length $\tau$.

We note that the backscatter coefficient can be considered in two parts, that due to molecules and that due to particulate matter in aerosols:

$$\beta(R, \lambda) = \beta_{\text{mol}}(R, \lambda) + \beta_{\text{aer}}(R, \lambda) \tag{15.12}$$

$\beta_{mol}$, mainly from oxygen and nitrogen, decreases with height above ground because of decreasing density. $\beta_{aer}$ varies in time and space and includes liquid and solid air pollution particles such as sulfates, soot, organic compounds, mineral dust, sea salt, pollen, rain drops, ice crystals, and hail [166].

### 15.2.4.4 *Transmission Loss*
Transmission loss $T(R, \lambda)$ is the fraction of light that is lost in two-way travel

$$T(R, \lambda) = \exp\left\{-2 \int_0^R \alpha(r, \lambda)\right\} dr \qquad (15.13)$$

The sum is extinction loss and $\alpha(r, \lambda)$ is the extinction coefficient. Similar to backscatter, equation (15.7), the extinction coefficient depends on the number concentration $N_j$ and the extinction cross section $\sigma_j$ of the $j$th scatterer:

$$\alpha(R, \lambda) = \sum_j N_j(R)\sigma_{j,ext}(\lambda) \qquad (15.14)$$

The extinction cross section $\sigma_{j,ext}(\lambda)$ can be partitioned into that due to scattering and that due to absorption and the extinction coefficient $\alpha(r, \lambda)$ can be additionally divided into that due to molecules and that due to aerosols.

### 15.2.4.5 *Lidar Equation*
From, these equations, we can write the lidar equation for the received signal power $P(R)$ in a lidar from a distance $R$ as

$$P(R) = KG(R)\beta(R)T(R)$$

$$= \eta_1 \frac{c\tau}{2} A \eta \frac{O(R)}{R^2} \beta(R, \lambda) \exp\left\{-2 \int_0^R \alpha(r, \lambda)\right\} dr \qquad (15.15)$$

where we used equation (15.4) for the system function $K$, equation (15.6) for the geometric $R$ dependence, $\beta(R, \lambda)$ for the unknown scattering coefficient, and equation (15.13) for the transmission loss. In addition to the lidar power in equation (15.15), there will be a background power that can be due to sun, stars, moon, and detector noise. The background power can be measured before emitting a pulse or at distances where no scatterers are expected. Background power is subtracted from the received lidar power before passing it to an optical spectrum analyzer for chemical analysis. The optical spectrum analyzer is described in Section 15.3 and its use in a lidar in Section 15.4.

## 15.3 SPECTROMETERS

Spectrometers or optical spectrum analyzers measure optical power as a function of wavelength [28, 29]. For absorption measurements, the molecular size of a particular chemical (such as $CO_2$) or bacteria (such as *Salmonella*) matches a specific wavelength of light, so that light at this wavelength is absorbed in causing resonance.

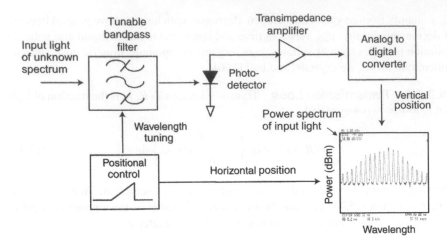

**FIGURE 15.2**    Generic laboratory optical spectrum analyzer.

## 15.3.1    Fabry–Perot-Based Laboratory Optical Spectrum Analyzer

Laboratory optical spectrum analyzers (OSAs) are used to view signals such as the spectrum of laser diodes and output of filters, but are not generally sensitive enough for determining chemical elements in a material or vapor. Fabry–Perot resonators (Section 6.2) may be used for spectrum analyzers by varying the spacing between two parallel mirrors. As one mirror moves with respect to the other, the half-wavelength that satisfies the boundary condition to fit between two mirrors determines the resonant wavelength. Such a tunable optical filter may be placed in Figure 15.2, a typical laboratory optical spectrum analyzer [29].

A sweep frequency causes the filter acting on the input light signal whose spectrum is sought to sweep linearly through wavelength. In a spectrum analyzer, each wavelength segment of the input is measured separately. As shown in Figure 15.2, an electric signal in the form of a sawtooth or ramp is used to move one mirror of a Fabry–Perot filter with a piezoelectric device to extract a wavelength band. The distance between mirrors determines the band center frequency and the reflectivity of the mirrors determines the bandwidth (Section 6.2). The electric signal for the sweep frequency is used to drive the horizontal axis of an oscilloscope as shown. The filtered light is detected by a reverse-biased photodetector and transimpedance amplifier [52] and the electric output fed to the vertical axis of the oscilloscope. On the screen we see the power spectrum of the input light, which is the intensity $(W/m^2)$ as a function of wavelength $(\lambda)$.

## 15.3.2    Diffraction-Based Spectrometer

### 15.3.2.1    *System for Diffraction-Based Spectrometer*    A glass prism (Figure 4.2b) will separate incoming light into its constituent colors but does not work at

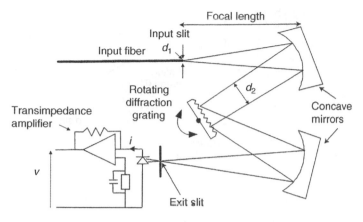

**FIGURE 15.3**   Grating spectroscope.

infrared, where many spectrometers operate: the size of chemical and bacteriological molecules often matches IR. Instead of a prism, a grating is commonly used (Figure 4.2c, Section 4.2), and this enables sufficiently high resolution for chemical and bacteriological detection.

Figure 15.3 [29] shows a diagram of a grating spectrometer. Light whose spectrum is sought enters through an input slit, width $d_1$, at left, and illuminates a curved mirror that collimates the light into a parallel beam of width $d_2$ to illuminate a grating. A curved spherical mirror is often used in preference to a lens because large mirrors are easier to manufacture. The collimated light is diffracted into multiple separate colors. The grating is rotated, normally with a stepping motor, which steps one color through to the output after another. Each of the colored beams is now focused in turn by the second spherical mirror onto an exit slit. In a monochromator, a single color of light will emerge from the exit slit. If a photodetector is placed behind the slit, an electronic signal shows the intensity of each waveband of light in a spectrometer. A transimpedance amplifier, also shown in Figure 15.2, is so named because it converts current $i$ proportional to light power into voltage $v$ with an operational amplifier as shown [52].

The optics should be diffraction limited (Section 3.4), which means that it is limited in performance only by diffraction. As described in Section 3.2.2, for an input slit of width $d_1$, the half-width of the main lobe is spatial frequency $f_s$ and angular spatial frequency $\omega_0$:

$$f_s = \frac{1}{d_1}\lambda f \quad \text{or} \quad w_0 = 2\pi \frac{1}{d_1}\lambda f \tag{15.16}$$

where $f$ is the focal length of the mirror and $\lambda f$ a scaling when taking a Fourier transform in optics with a mirror of focal length $f$ using Fraunhofer approximation (Section 3.3.5) with $z = f$. Similarly, for focusing to the output slit, the collimated beam of width $d_2$ is focused on the exit slit. The light focused on an exit slit must have a main lobe width greater than the exit slit width for maximum signal.

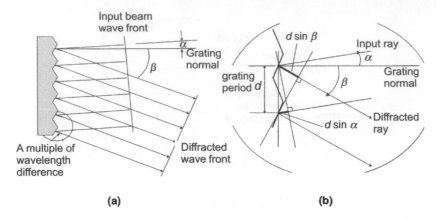

**FIGURE 15.4**  Bending light with blazed grating: (a) overall grating and (b) expanded period of grating.

### 15.3.3  Grating Operation in Spectrometer

The refractive grating diffracts light (Section 4.2), bending it according to wavelength, $\lambda$. It has a substrate, periodic perturbations, and a refractive coating. We derived the grating equation in Section 4.2.1. Here we show its validity for a blazed grating [29]—one to emphasize a single order of diffraction. Figure 15.4a shows an idealized representation of an input beam of specific wavelength diffracted by reflection from a blazed grating. From Figure 15.4b, compared to a normal incident ray, the upper edge of the collimated input beam travels a distance $b$ farther to reach the marked output wave front, while the lower edge travels a distance $a$ farther. If $d$ is the width of the collimated beam, and $\alpha$ and $\beta$ are the angles for input and output beams, $a = d \sin \alpha$ and $b = d \sin \beta$, as labeled in Figure 15.4b. Hence, the difference in distance traveled for the upper and lower edges of the collimated beam is

$$b - a = d(\sin \beta - \sin \alpha) \tag{15.17}$$

For the upper and lower edges of the collimated beam to contribute to a wave front, the difference in length must be a whole number $n$ of wavelengths $\lambda$,

$$d(\sin \beta - \sin \alpha) = n\lambda \tag{15.18}$$

This is identical to the grating equation derived for any grating, equation (4.1), except that the sign is changed here because we have a reflective grating rather than a transmitting one.

When light returns along the same path as the input $-\beta = \alpha = \theta$, we have

$$n\lambda = 2d \sin \theta \tag{15.19}$$

known as the Littrow condition, illustrated in a device in Figure 15.5 [53]. An incoming beam passes through a slit in the grating and diffracts out as ray 1 to illuminate a collimating mirror. For a specific wavelength, the collimated light beam, rays 2,

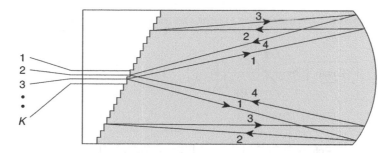

**FIGURE 15.5**   Device using Littrow condition grating.

returns to the grating and the beam is diffracted by the grating along rays 3 and returns to the collimating mirror that now focuses the light to a spot on the grating via rays 4. Each color focuses to a different position as shown at the different λ values out at left.

***15.3.3.1  Limits to Grating Resolution***   The grating imposes a limit to the wavelength resolution, how finely colors may be separated. The blazed grating (Figure 15.4) has length $Nd$, where $N$ is the number of grating periods and $d$ is the grating period. Like an antenna, a longer grating produces a narrower beam pattern. For exit angle $\beta$, the 3 dB width at which power emitted is half the peak is (equation (3.20), [29])

$$\Delta \beta_m = \frac{\lambda}{Nd \, \cos \, \beta} \tag{15.20}$$

Wavelength resolution is further limited by the dispersion equation (15.18). Taking a derivative $d\beta/\Delta\lambda$ of equation (15.18), we can write

$$\Delta \beta = \frac{n}{d \, \cos \, \beta} \Delta \lambda \tag{15.21}$$

where $n$ is the order of the grating.

For minimum wavelength resolution, $\Delta\lambda_{min}$, equating (15.21) and (15.20) to eliminate $\Delta\beta S$ and solving for $\Delta\lambda_{min}$ shows resolution depends only on grating length and $\lambda$

$$\Delta \lambda_{min} = \frac{\lambda}{Nn} \tag{15.22}$$

Note that using a different focal length lens or mirror for focusing to the output affects magnification but not the resolution.

### 15.3.4   Grating Efficiency

The efficiency with which input light is used for producing output depends on the quality of the grating, both the quality of the coating and the quality of the blaze,

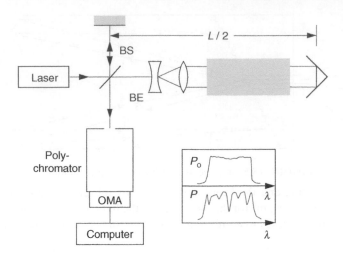

**FIGURE 15.6**  Detecting chemical and biological weapons by transmission.

which emphasizes order $n$ at the expense of other orders. As reflection is polarization sensitive, a less preferable polarization direction will reduce efficiency.

The output is focused on an exit slit; the narrower the slit, the smaller the part of the spectrum selected, and therefore the higher the spectral resolution. But the highest resolution remains limited by the grating resolution (15.22). A spectrum analyzer has a photodetector after the exit slit. The type of photodetector depends on wavelength and sensitivity. For high-resolution spectroscopy, a photomultiplier is used. If high resolution is not required, a photodiode is adequate.

## 15.4  SPECTROSCOPIC LIDAR SENSES CHEMICAL WEAPONS

Two approaches to sensing chemical and bacteriological elements in the battlefield or terrorist environments are by direct transmission or backscatter [28].

### 15.4.1  Transmission Detection of Chemical and Biological Materials

Direct transmission is normally performed close to the ground. Figure 15.6 [28] shows a transmission optical spectrometer in which laser light from a tunable laser diode or other fixed laser that covers several molecular sizes of interest is passed to a beam splitter BS. One output of the beam splitter feeds a beam expander BE, or inverting telescope (Section 1.3.2), to produce a collimated beam passing through the medium (shaded) in which chemical or biological weapons material may exist. The other output from the beam splitter is reflected by a mirror at top into the polychromator as a reference for the power of the laser $P_0$. A polychromator is a spectrum analyzer that separates the light into constituent wavelengths.

A retroreflector reflects the light reduced by absorption of power $P$ back to the beam splitter BS that passes it to the polychromator. Each wavelength passes to a different channel of an optical multichannel analyzer (OMA). A plot of power versus wavelength is obtained from the computer. If a specific molecule from a weapons chemical or bacteria is present, the wavelength matching the molecular size will experience attenuation due to the power absorbed by resonance. Several different molecule types are detected simultaneously. The fraction of the input power $P_0$ received for a given frequency after traveling a distance $L$ is

$$P(L) = P_0 \exp\{-a(\omega)L\} \tag{15.23}$$

where the attenuation coefficient $\alpha(\omega) = N_i\sigma_{i,\text{abs}}$ depends on the concentration $N_i$ of absorbing molecules with absorption cross section $\sigma_{i,\text{abs}}$. An attenuation coefficient $a(\omega)$ may be written as

$$a(\omega) = \alpha(\omega) + S \tag{15.24}$$

where the sum of absorption from all particles is

$$\alpha(\omega) = N_k\sigma_{k,\text{abs}} \tag{15.25}$$

and the sum of the scattering from all particles $S$ is

$$S = \sum N_k\sigma_{k,\text{scat}} \tag{15.26}$$

$S$ is mainly Mie scattering from particles, dust, and water drops. A small part is from Rayleigh scattering from molecules and atoms for which the wavelength is greater than the molecule size. $S$ has a broad spectrum while $\alpha(\omega)$ has a very narrow linewidth (a few GHz) matching a specific molecule. Consequently, sensitive detection is obtained by comparing the input power at the exact frequency for the peak power for a specific chemical molecule $P(\omega_1, L)$ with the output power for neighboring frequencies that change little with frequency $P(\omega_2, L)$. The difference values are simultaneously computed for a number of known chemical and biological agents. Detection depends on the ratio

$$\frac{P(\omega_1, L)}{P(\omega_2, L)} \approx \exp\{-N_i[\sigma_{i,\text{abs}}(\omega_1) - \sigma_{i,\text{abs}}(\omega_2)]L\} \tag{15.27}$$

### 15.4.2 Scattering Detection of Chemical and Bacteriological Weapons Using Lidar

As described in Figure 15.1 (Section 15.2), pulsed laser light moves through a telescope (Section 1.3.4) to form a wide beam that is projected toward a questionable cloud in the sky to determine its chemical composition. The light scattered from the cloud $S(\lambda, t)$ returns to a photomultiplier tube (PM) and is measured in wavelength and in time [28]. The scattered light is gated during the time interval $t_1 \pm \Delta t$ to determine the time signal at $t_1 = 2R/c$ (equation (15.1), Section 15.2.4.1).

$$S(\lambda, t_1) = \int_{t_1 - \frac{1}{2}\Delta t}^{t_1 + \frac{1}{2}\Delta t} S(\lambda, t)dt \qquad (15.28)$$

The magnitude of $S(\lambda, t)$ depends on the solid angle $d\Omega = D^2/R^2$ covered by the telescope, diameter $D$, and aperture $A = \pi(D/2)^2$; the concentration of the molecules signal $N$; and backscattering cross section $\sigma_{scat}$. We can write

$$S(\lambda, t) = P_0(\lambda) \exp\left\{-2 \int_0^R \alpha(r, \lambda)\right\} dr \, N\sigma_{scat}(\lambda)D^2/R^2 \qquad (15.29)$$

This is similar to the lidar equation (15.15) with 100% overlap, $O(R) = 1$, efficiency $\eta = 1$, signal for a range bin $c\tau/2$, antenna perception $A/R^2 \approx D^2/R^2$, and scattering coefficient $\beta = N\sigma_{scat}$.

Similar to the transmission approach (Section 15.4.1), we consider a wavelength $\lambda_1$ at an absorption line (the resonance peak) of a molecule and a wavelength $\lambda_2$ where absorption is negligible. For small $\Delta\lambda = \lambda_1 - \lambda_2$, the variation in scattering cross section may be neglected. Then, ratio of scattering for $\lambda_1$ and $\lambda_2$ is

$$\frac{S(\lambda_1, t)}{S(\lambda_2, t)} = \exp\left\{2 \int_0^R [\alpha(\lambda_2) - \alpha(\lambda_1)]dR\right\}$$

$$\approx \exp\left\{2 \int_0^R N_i(R)\sigma(\lambda_1)dR\right\} \qquad (15.30)$$

Equation (15.30) provides the concentration of molecules sought $N_i(R)$ integrated over the complete path to cloud and back.

The variation in concentration of molecules with distance is obtained by sampling alternate $\lambda$ values at successive intervals of $\Delta t$. Then absorption for a range bin at distance $R$ is obtained (using a series expansion for the exponential) by

$$\frac{S(\lambda_1, t)/S(\lambda_2, t + \Delta t)}{S(\lambda_1, t)/S(\lambda_2, t)} = \exp\{-[\alpha(\lambda_2) - \alpha(\lambda_1)]\Delta R\}$$

$$\approx 1 - [\alpha(\lambda_2) - \alpha(\lambda_1)]\Delta R \qquad (15.31)$$

According to equation (15.25), the concentration of molecules, $N_i$ for the $i$th chemical element, can be obtained from the absorption coefficients in equation (15.31). The concentration of a toxic chemical or bacteria is used to assess the nature of a threat.

# CHAPTER 16

# 94 GHz RADAR DETECTS/TRACKS/
# IDENTIFIES OBJECTS IN
# BAD WEATHER

At 94 GHz the radar uses many of the light technologies described in Part I because the frequency falls between light and microwaves. For example, the gyroklystron used (Section 10.1) has a quasi-optical mirror resonator and the 94 GHz radar uses a quasi-optical duplexer to separate transmit and receive waves. This frequency is considered the upper part of the millimeter band of frequencies and has distinct advantages over light or microwaves in several applications. The frequencies classified as W-band, 75–110 GHz, have wavelengths of a few millimeters (4–2.7 mm), which allows them to pass through particles of smaller size. Of particular interest is $f -$ 94 GHz that falls in an atmospheric window and has a wavelength in air of $\lambda = c/f = $ 3.2 mm, where $c = 3 \times 10^8$. We include consideration of W-band in this book because electromagnetic waves between light and microwaves can be generated efficiently at high power using free electron lasers or other cyclotron resonators or amplifiers (Chapter 10) and because resolution and visibility are better than conventional radar, especially at 94 GHz, the atmospheric window.

In Section 16.1, we discuss propagation of electromagnetic radiation through the atmosphere and show the window at 94 GHz. In Section 16.2, we describe the design of the high-resolution, high-power 94 GHz W-band radar developed at Navy Research Laboratories. In Section 16.3, we describe four applications: monitoring satellites in

*Military Laser Technology for Defense: Technology for Revolutionizing 21st Century Warfare,*
First Edition. By Alastair D. McAulay.

low earth orbit, recording and monitoring space debris, detecting and identifying moving objects by Doppler sensitivity, and low elevation naval operations.

## 16.1    PROPAGATION OF ELECTROMAGNETIC RADIATION THROUGH ATMOSPHERE

At the atmospheric window of 94 GHz, the wavelength of 3.2 mm is between millimeter microwaves and optics. In contrast to light, 94 GHz passes through most bad weather such as fog, smog, dust, snow, modest rain, and clouds. At the same time, the resolution is better than that for microwaves. This has tremendous value in the battlefield for high-resolution imaging, communications, crowd control, weather assessment, and radar—the topic of this chapter. Attenuation of electromagnetic waves through the atmosphere at microwave and optical frequencies is shown in Figure 16.1 [66]. Alternatives and development of this graph are described in Ref. [5]. For fog of $0.1 \text{ g/m}^3$ corresponding to visibility of 50 m, the dashed line at upper right in Figure 16.1, visible and infrared light (labeled on $x$ axis) have unacceptably high attenuation between 100 and 210 dB/km. In contrast, on a clear day, attenuation of visible light drops from 209 to 0.02 dB/km at bottom right. From 10 THz (30 μm) to 1 THz (0.3 mm), the average attenuation in air is very high at over 100 dB/km.

At 94 GHz, around 3 mm, a significant dip in attenuation signifies an atmospheric window that we can exploit in military applications. This window falls in W-band, labeled along the bottom left of the graph. The 94 GHz dip falls between oxygen resonances. The attenuation of a 94 GHz W-band wave in $0.1 \text{ g/m}^3$ fog is 0.4 dB/km,

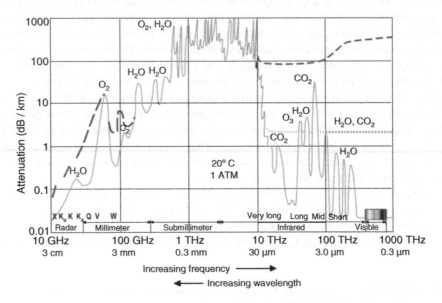

**FIGURE 16.1**    Propagation through the atmosphere.

only slightly more than that without the fog on a clear day, 0.33 dB/km. The ability to see through fog and dust is valuable for military applications. For drizzle of 0.25 mm/h, the attenuation at 94 GHz is virtually unseen and comparable with visible light in the absence of drizzle. For heavy rain at 25 mm/h, the attenuations are 12 and 9 dB/km for W-band and visible light, respectively.

The attenuation at left decreases for the radar bands for which wavelength exceeds particle size but the resolution decreases to 3 cm at X-band. The heavy dashed line at left represents attenuation for heavy rain. In summary, if we need resolution of around 9 mm, $K_a$ band has low attenuation of 0.13 dB/km through the atmosphere; however, if we need three times better resolution, approaching 3 mm, W-band is suitable but has slightly more attenuation at 0.29 dB/km.

## 16.2  HIGH-RESOLUTION INCLEMENT WEATHER 94 GHz RADAR

### 16.2.1  94 GHz Radar System Description

A 94 GHz radar [74] was developed at Naval Research Laboratories. One planned application was for tracking and imaging of space debris [158, 159]. Such a high-power gyrotron was proposed earlier at NRL for a super range resolution radar and for atmospheric sensing [76]. The description for this radar follows that in Ref. [74]. The operating parameters are provided in Table 16.1 [74].

The operating modes, waveforms, and signal characteristics are provided in Table 16.2 [74]. Optimum waveforms for each mode are synthesized electronically and then modulate light at 94 GHz that is amplified for transmission. The search mode uses the whole range interval as distance to the target is unknown and analog pulse compression used. Once distance is known, tracking uses the limited range interval for this distance. Digital pulse compression is used [72]. For target identification, we use a wideband signal (<600 MHz) and stretched pulses (<100 μs) to obtain detailed backscatter for identification. Remote sensing mode enables investigation of cloud

TABLE 16.1  Operating Parameters of
94 GHz Radar [74]

| Radar Parameter | Value |
| --- | --- |
| Frequency | 94.2 GHz |
| Bandwidth | 600 MHz |
| Peak power | 70 kW |
| Average power | up to 7 kW |
| Antenna diameter | 2 m |
| Antenna gain | 62.5 dB |
| Polarization | Circular or linear |
| Transmit loss | 3.5 dB |
| Receive loss | 3.0 dB |
| Noise figure | 8 dB |

**TABLE 16.2**    **Operating Modes for 94 GHz Radar**

| Mode | Pulse Width ($\mu s$) | Band width (MHz) | PRF (kHz) | Burst length (No. of Pulses) |
|---|---|---|---|---|
| Search/acquisition | 17 | 14 | 5 | 64 |
| Track | 20 | 18 | 5 | 128 |
| Stretch | 20–100 | $\leq 600$ | 1–5 | 100–1000 |
| Remote sensing | 0.1–1 | 1–10 | 5 | 1–100 |

physics for weather assessment. More pulses in the burst length increase the time length of the signal for better frequency resolution.

A block diagram of the 94 GHz radar system (Figure 16.2) [74] has transmit, antenna, and receive parts. Waveforms for transmission are developed in electronic waveform synthesizers at 60 MHz and upconverted through 1.8 and 10–94 GHz. The wideband signal is generated at 255 MHz and upconverted appropriately. The final conversion to 94 GHz is performed with a 94 GHz local oscillator and mixer at upper left. The operating mode is selected at the mixer that converts to 94 GHz. The signal feeds to a commercially available ordinary (O-type) traveling-wave tube (TWT) [24] that amplifies the signal to 50 W (Section 10.1, Figure 10.1). The output of the TWT feeds into a 100 kW gyroklystron amplifier built at Naval Research Laboratories (Section 10.1, Figure 10.1) [65], described next.

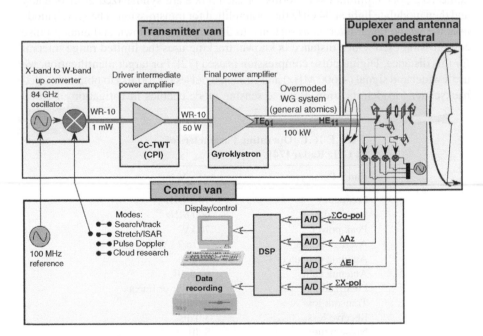

**FIGURE 16.2**    Block diagram of 94 GHz radar system.

The received 94 GHz signal is converted to X-band (10 GHz) with a second 84 GHz local oscillator at the antenna. The two 84 GHz local oscillators, one in transmit and the other in receive, are phase locked to a single 100 MHz reference to create a totally phase coherent system. The received signal passes through downconversion to 1.8 GHz and 60 MHz (opposite to the upconversion in the transmitter).

## 16.2.2 Gyroklystron with Quasi-Optical Resonator

A klystron amplifier is fed with an electron beam, nonrelativistic for an O-type conventional [24] tube and relativistic for the much more powerful gyroklystron amplifier (Figure 10.1, Chapter 10). The electron beam, in passing from the cathode gun to an anode, moves through two cavities (or more in some klystrons): the input signal to be amplified enters the first or buncher cavity and the second cavity amplifies the signal and provides the amplified output.

Because 94 GHz is between light and microwaves, the klystron resonator may be quasi-optical. To clarify the term quasi-optical resonator, a conventional single-cavity gyrotron resonator, Figure 16.3a (Figure 10.5), is compared with a single-cavity quasi-optical resonator in Figure 16.3b [65]. In a conventional microwave gyrotron (Figure 16.3a), the tube along which the electron beam passes is blocked at the ends to form a vertical cylindrical cavity. The 94 GHz radiation generated passes out of the top end of the cylindrical cavity. In a quasi-optical cavity (Figure 16.3b), Fabry–Perot type mirrors (Section 6.2) are placed at right angles to the electron beam path and the 94 GHz radiation exits through left or right mirrors like a laser.

The quasi-optical resonator schematic for an early version of the two-cavity NRL gyroklystron is shown in Figure 16.4 [40]. In comparison with the conventional cylindrical cavity, the quasi-optical resonator has narrower bandwidth, is simpler

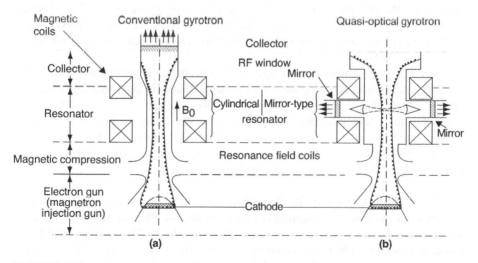

**FIGURE 16.3** Comparison of conventional and quasi-optical resonators in klystron: (a) conventional cylindrical cavity and (b) quasi-optical cavity.

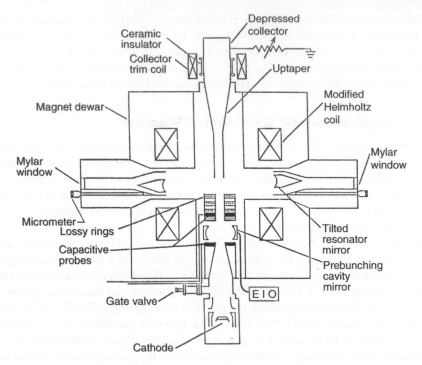

**FIGURE 16.4**   Schematic for early gyroklystron.

to construct, has better isolation, and is more adaptable, but may be less efficient. The electron beam moves upward from the cathode to the collector. Prebunching mirrors are used for the first transverse cavity of the two-cavity gyroklystron. The second cavity is transverse across the center with Mylar windows. The square boxes with the diagonal crosses are the magnetic coils that provide the cyclotron magnetic field (Section 10.2.2.1).

Figure 16.5 shows for the two-cavity NRL gyroklystron, (a) photograph of an early prototype with superconducting magnet at center [40] and (b) photograph of a recent version without the superconducting magnet [74]. In Figure 16.5a, the cryogenic system for the superconducting magnets is seen to be bulky. In Figure 16.5b, the electron gun shoots relativistic electrons upward into the main cavity. The first nozzle out to the right is the input for the signal to be amplified and the next nozzle is for the amplified signal output. The magnets are discussed next.

### 16.2.2.1   *Magnetic Field for Gyroklystron*   In a gyroklystron, the magnetic field provides the essential coupling between emitted 94 GHz RF field and the cyclotron beam modes on the electron beam (Section 10.2.1). The magnetic field required, $B$, is computed from the frequency $f = 94$ GHz by $B = f/(28 \times s) = 3.36$ T, where $s$ is the harmonic number ($s = 1$ for the gyroklystron used). To reduce weight and size requirement, a superconducting magnet was used with niobium–titanium

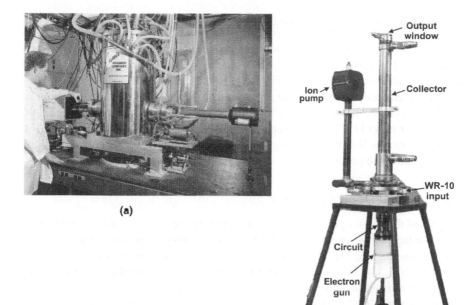

**FIGURE 16.5** Photographs of NRL gyroklystron: (a) prototype with cryogenic superconducting magnet and (b) that without superconducting magnet.

wire at 9 K. A closed-cycle cryocooling system provides 1 W cooling at 4 K. The magnet takes 20 min to reach its full field after which it takes relatively little power because of the low superconductor resistance.

***16.2.2.2 Modulator*** The anode of the electron gun of the gyroklystron is modulated at a maximum peak repetition frequency of 5 kHz with an 18 kV switch. The pulse width is 100 μs and maximum duty cycle is 10%. The cathode voltage of the gun is < 70 kV and cathode reference is 30 kV.

### 16.2.3 Overmoded Low 94 GHz Loss Transmission Line from Gyroklystron to Antenna

Because of the high multi-kW power at 94 GHz and the duplex nature of radar, the transmitter equipment cannot be mounted directly on the antenna as in Section 17.1, where no return signal is received. Therefore, we require a low-loss transmission line and rotary joint to carry transmitted power from the final power amplifier to the antenna-mounted duplexer. A conventional 94 GHz waveguide suffers 3 dB/m loss at these high peak and average powers. A circular overmoded waveguide is used in which the waveguide transverse dimension, diameter 32 mm, is many times the wavelength, for excess modes (Figure 16.2). Then losses for a smooth-walled waveguide are below 0.01 dB/m for the $TE_{01}$ mode and less for the $HE_{11}$ mode [7]. The gyroklystron emits a $TE_{01}$ mode into the circular waveguide, which is then converted with a curved corrugated waveguide into an $HE_{11}$ mode for coupling to the rotating antenna (Figure 16.2). The $HE_{11}$ mode emits a Gaussian beam into free space as is required for input to the quasi-optical duplexer attached to the antenna. The $HE_{11}$ mode is converted to circular polarization with a grating polarizer built into a high-power miter [73, 149, 172] (angled) bend. Circular polarization is required so that polarization will remain unchanged as the antenna rotates. After passing through the rotary joint, the polarization is converted back to linear for the duplexer.

### 16.2.4 Quasi-Optical Duplexer

The goal of the quasi-optical duplexer is to separate the transmit and receive beams. The advantage of the quasi-optical duplexer is low loss, high power handling, and broad bandwidth [38, 74]. The linear polarized $HE_{11}$ mode from the transmission line in Section 16.2.3 launches a linear polarized $TEM_{00}$ mode into the input port of the duplexer at the right-side of Figure 16.6 [74]. Clamshell reflectors direct the beam onto the first wire grid polarizer, where parallel wires allow a polarized signal, with $E$ field oriented along spaces, to pass. For light propagating from right to left, the wires are at $-45°$ as shown; that is, the field to the left of the wire polarizer is linearly polarized at $-45°$. The Faraday rotator rotates the polarization clockwise through $45°$ to become vertically or zero polarized and passes it on to the second wire grid polarizer through which it passes because the second wire polarizer is also oriented vertically, at $45°$ to the first wire grid polarizer. The Faraday rotator is commonly used for isolators [176]. For the high power and frequency involved, isolators are preferable to quarter-wave plates (Section 2.2.1.2). In Figure 16.6, the signal passing through the vertical wire polarizer couples to the Cassegrain antenna (Section 1.3.4.1).

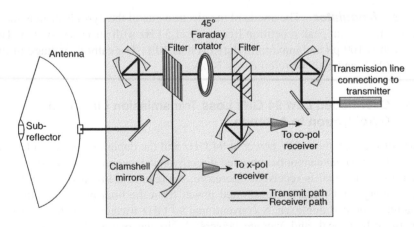

**FIGURE 16.6**  Quasi-optical duplexer.

On return from the target, at the output of the Cassegrain antenna, the cross-polarized component relative to that transmitted from the Cassegrain reflects off the vertical wire grid polarizer (as if a polarized beam splitter) and feeds to the cross-polarization receiver, the lowest receiver in the control van in Figure 16.2. In contrast, the copolarized signal, polarized at right angles to the cross-polarized signal, passes through the vertical wire grid polarizer. The zero polarized signal receives another 45° clockwise rotation while traveling left to right through the Faraday rotator so that it is polarized at right angles to the right side or −45° wire polarizer and bounces off it and is directed to the input of the copolarization receiver, the uppermost of the receivers in the control van in Figure 16.2. Separating the polarizations into cross- and copolarizations at the receiver provides additional information relating to propagation and the nature of the target. The cooling system for the Faraday rotator had to be redesigned to handle 100 kW power.

Measurements from testing are described in Ref. [74]. By reflecting all the radiation back into the duplexer with an aluminum plate, a two-way loss of 4.8 dB was measured for the duplexer. By replacing the aluminum sheet reflector with absorbing material, the amount reflected back inside the transmitter was −17 dB. The leakage or loss in passing through the antenna was measured as −30 dB. Frequency response for combined gyroklystron amplifier, overmoded transmission line, quasi-optical multiplexer, and antenna was measured in a field test over 700 m. For a linear frequency modulation [141] and a 50 μs chirp pulse of bandwidth 400 MHz centered at 94 GHz, the resulting pulse was flat, within 1.5 dB, over the 400 MHz bandwidth of the chirp bandwidth.

## 16.2.5  Antenna

A symmetric Cassegrain antenna was used (Section 1.3.4.1) for which the feed is lined up with the axis of the antenna. This provides a stiff structure because the receiver and duplexer, which feeds the antenna, are mounted directly on the antenna. This minimizes losses and reduces aperture losses compared to placing a deflecting mirror in the antenna output path, the Nasmyth telescope (Section 1.3.4.2, Figure 15.1). The overmoded waveguides and miter bends can be seen in the photograph of the back of the antenna (Figure 16.7) [74]. Two further miter bends are used to align the waveguide with the elevation axis of rotation before the elevation rotary coupler and a miter bend directs power to the duplexer box. The duplexer box emits at the center of the antenna to illuminate the subreflector of the 1.8 m Cassegrain bidirectional antenna, machined to high precision for operation in a variable environment. Measurements showed a beamwidth of 0.11° and an antenna gain (peak over omnidirectional response) of 62.7 dB.

## 16.2.6  Data Processing and Performance

The data processing setup is described in Ref. [74]. The performance of the complete system was measured in terms of range in the earth's atmosphere. The range is substantially larger with altitude because turbulence decreases. Using the parameters in Table 16.2 and a 100 μs pulse, the signal to noise ratio on a 1 m$^2$ target at 1000 km

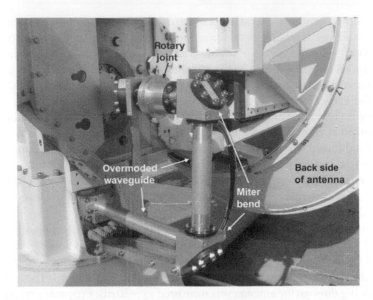

**FIGURE 16.7** Back of antenna showing overmoded waveguides and miter bends.

was −1 dB. For imaging with a 100 ms dwell, 30° elevation, and 20 dB signal to noise ratio, the range was 680 km. For tracking with 60 ms dwell, 30° elevation, and 20 dB signal to noise ratio, the range was 570 km. Low elevations are worse because the beam spends more time close to the earth, where turbulence is highest. Calibration measurements are described in Ref. [74] and depend on water content in the air (1.4 dB/km at 9 g/m$^3$).

## 16.3 APPLICATIONS, MONITORING SPACE, HIGH DOPPLER, AND LOW SEA ELEVATION

Four applications for the 94 GHz radar are discussed.

### 16.3.1 Monitoring Satellites in Low Earth Orbit

Since World War II, the advantage is with the side that controls the air space above the conflict area. In future conflicts, the satellite space (lower earth orbit) will be equally if not more critical because we rely on many different types of satellites both for commercial activity and in a conflict. We rely on GPS satellites to keep track of our forces and those of our allies and to steer our missiles; FedEx would be lost without it, not to mention many vehicle drivers. We also rely on communication satellites to communicate across the battlefield, not to mention their use in credit card and ATM pin verification. Our imaging satellites keep surveillance (every 90 min) around most of the critical areas of the globe and have sufficient resolution to see humans and address environmental issues. One country has reportedly used dazzle lasers to blind a U.S.

**FIGURE 16.8**    Illustration of debris in lower earth orbit.

imaging satellite; this could portend another cat and mouse laser-related technology contest. Lasers are highly effective in space because of their narrow beams and because there is virtually no bad weather to block the light. However, we frequently need to monitor satellite space from the ground or a lower altitude where the atmosphere restricts the use of light. The 94 GHz quasi-optical radar described in this chapter is ideal for this because 94 GHz falls in a low attenuation atmospheric window and because it is between light and microwaves, that is, has high enough frequency to image small particles and low enough frequency to pass through inclement weather.

## 16.3.2    Problem of Detecting and Tracking Lower Earth Orbit Debris

Unwanted debris in lower earth orbit, altitude 200–3850 km, consists of over 300,000 items larger than 1 cm, 21,000 over 10 cm, and thousands over 2 kg (Figure 16.8) [137]. In addition, there are over 800 active satellites that can collide with other satellites. Debris traveling at 5–10 km/s can damage satellite solar cells and present a threat to astronauts working outside their spacecraft. The level of debris is increasing because of the growth in strategic importance of space and because it is unclear who is responsible for cleanup of this region. If debris continues to build up, it may make it very difficult for humans to leave earth in a few years.

In 2001, radar capable of finding and tracking 1 cm sized debris took a significant step forward when the power of 94 GHz sources increased by 20 times to 10 kW average (Chapter 10). Higher power provides greater range; 94 GHz falls in an atmospheric window and passes through inclement weather (Figure 16.1, Section 16.1) and provides less than 1 cm resolution. A 94 GHz radar is required to pass through the atmosphere and provide adequately high resolution in combined parameter space of range, velocity, and angle.

### 16.3.3 Doppler Detection and Identification

The 94 GHz radar can be used to identify objects because the Doppler sensitivity is higher than that for conventional microwave radar because of its higher frequency. Doppler versus range pictures for inverse synthetic aperture radar (ISAR) for target identification are shown in Ref. [74]. Doppler frequency $f_d$ increases with frequency $f$: $f_d = f(1 + (u/c) \cos \theta)$, where $u$ is the velocity of motion, $c$ is the speed of light, and $\theta$ is angle between direction of velocity and line to the target. So at $f = 94$ GHz, the Doppler frequency is 10 times larger than that at 10 GHz and this provides good Doppler images for identification even when there is only small motion. For example, the vibration of a small plane or slow turns provide adequate Doppler for identification.

### 16.3.4 Low Elevation Radar at Sea

Another application is shown in Ref. [74] that reduces a problem with conventional microwave radar for low-angle Navy tracking caused by multipaths bouncing off the sea surface that confuse the tracker. The tracker sees a second image of a target in the water. W-band at 94 GHz reduces this problem because it has much narrower beamwidth and the shorter wavelength reduces specular reflection from the surface variations on the ocean.

# CHAPTER 17

# PROTECTING FROM TERRORISTS WITH W-BAND

Several useful applications are discussed in which W-band electromagnetic radiation is used to protect from terrorists. W band covers the frequencies from 75 to 110 GHz with corresponding wavelengths of a few millimeters (4–2.7 mm) (Section 16.1) [5]. The higher frequency end of W-band is approaching optics so that some quasi-optical techniques may be used, such as mirror resonators (Section 16.2.2, Figure 16.3) and quasi-optical duplexer (Section 16.2.4). Moreover, in Section 16.1, Figure 16.1 [5], W-band provides an application trade-off between attenuation through the atmosphere and resolution. For example, at higher frequencies than W band, left side of Figure 16.1, wavelength and resolution decrease but attenuation increases. As discussed in Chapter 10, W-band frequencies can be generated efficiently at very high power using free electron lasers or other cyclotron resonators or amplifiers. We can also generate high peak power in this frequency range with explosives (Section 17.4.1.1) [3, 123].

In Section 17.1, we discuss a system, called Active Denial, that uses W-band to break up crowds in which terrorists may be hiding with concealed weapons. The technology is similar to the transmit side of the 94 GHz radar described in Chapter 16, so the details are not repeated in this chapter. In Section 17.2, we discuss nonionizing airport body scanners that are safer than X-rays for uncovering concealed weapons. In Section 17.3, we inspect unopened packages with W-band radiation that passes through Styrofoam and cardboard but reflects or scatters from the package contents. In Section 17.4, we discuss the use of W-band to interfere with or destroy enemy

*Military Laser Technology for Defense: Technology for Revolutionizing 21st Century Warfare*,
First Edition. By Alastair D. McAulay.
© 2011 John Wiley & Sons, Inc. Published 2011 by John Wiley & Sons, Inc.

electronics. We also discuss methods of protecting our electronic equipment from electromagnetic radiation used by enemy.

## 17.1 NONLETHAL CROWD CONTROL WITH ACTIVE DENIAL SYSTEM

Terrorists carrying concealed weapons can hide in a crowd. As most of the people in a crowd are likely to be innocent civilians, it is better to disperse the crowd than attack it or threaten it by shooting live or rubber bullets at them or over their heads. A crowd may be dispersed rapidly with electromagnetic radiation at W-band in systems such as Active Denial (Figure 17.1) [10, 139].

94 GHz, at the high end of W-band, falls in a dip in Figure 16.1 (Section 16.1), which is in an atmospheric window. Because it passes through dielectric particles smaller than its wavelength (3.2 mm), 94 GHz will pass through varied weather, fog, smog, dust, snow, modest rain, and clouds. On reaching a person, it passes through clothes, but reflects from skin. Conducting clothes can be worn to block it. In the process, with adequate power, the 94 GHz electromagnetic wave heats up water and

**FIGURE 17.1**  Active denial system on a Humvee.

fat molecules near or on the surface of the skin, causing perspiration to boil. At the higher W-band frequency of 94 GHz, the wave penetrates less than 0.04 mm into the skin, thus avoiding nerve endings and blood vessels that could otherwise produce permanent damage. A microwave oven at 2.45 GHz penetrates deeper and is more dangerous. The rising temperature creates intense pain causing those exposed to instantly flee from the radiation. Eyelids are most vulnerable because they have only a thin top layer of skin, but people naturally protect their eyelids by turning their heads away at the first sign of pain.

With typical power of 2.5 MW, the beam can be used at distances up to 700 m and penetrates thick clothes. The beam is narrow at 94 GHz. The tests suggest that there is only 0.1% chance of injury. Eye glasses and body piercings are not affected but tattoos can release toxins that make people feel sick. The system was not used in Iraq, perhaps because of operational constraints regarding overuse or critical intensity setting. Raytheon has developed a commercial system.

The generation of 94 GHz at high power is described in Chapters 10 and 16. Because active denial involves only transmit and not receive part, the active denial system differs from the 94 GHz radar in Chapter 16 by omitting the receive part of the system. Because of reduced size and weight, the transmitter is now mounted directly on the Cassegrain antenna (Section 1.3.4.1) and the system is small enough to fit on a Humvee (Figure 17.1). The large dish on the roof is the main mirror of the Cassegrain telescope, used as a beam expander or inverse telescope, and the small convex mirror of the Cassegrain telescope sticks out over the windshield.

## 17.2 BODY SCANNING FOR HIDDEN WEAPONS

As mentioned in Section 17.1, because millimeter waves pass through dielectric particles smaller than their wavelength, they pass through most clothes and reflect or scatter from skin. This is used in full body scanners to disclose concealed weapons at a standoff distance of up to 100 m [5]. Less power is needed for body scanning than for active denial (Section 17.1). An existing 35 GHz $K_a$ band radar was adapted for demonstration in early scanners. An example of an image produced at 33 GHz is shown in Figure 17.2a [150] in which a concealed handgun shows up tucked into a waist band. Concealed plastic explosives such as C4 also show up on 33 GHz scans [5].

Frequencies higher than 94 GHz have been used in experiments. The shorter wavelength gives higher resolution but is attenuated more, which reduces the standoff distance for detection of concealed weapons (Section 16.1) [5]. Many researchers are also experimenting with passive millimeter wave body scanning, for which there is no source but a 94 GHz filter is placed in front of the receiver. The body naturally emits electromagnetic radiation and there is a difference in emissivity and reflectance between skin and explosives or metal. However, results vary with environmental factors such as temperature. Figure 17.2b [5] shows a 94 GHz passive image of a man in the open carrying a knife hidden in a newspaper. The newspaper appears transparent at 94 GHz.

**FIGURE 17.2** Millimeter wave full body scans of man: (a) active 33 GHz, concealed gun at waist and one in hand, and (b) passive 94 GHz, carrying knife hidden in a newspaper.

Millimeter waves compete with X-ray body scanners that seem to have been adapted from airport baggage scanners. Figure 17.3a shows an X-ray of a dressed man carrying a concealed gun and package at his front and Figure 17.3b shows the X-ray of his back, where he carries another concealed package at the small of his back and a knife at his ankle. The photon energy E of X-rays is much higher than that for millimeter wave photons according to Einstein's light–matter interaction theory $E = h\nu$ (Section 7.1.2), where $h$ is Plank's constant and $\nu$ is frequency. The frequency for X-rays is over $3 \times 10^{16}$ compared to $100 \times 10^{9}$ for millimeter waves at 100 GHz. Thus, the X-ray frequency is 300,000 times higher than the millimeter wave frequency, which means that an X-ray photon has 300,000 times as much energy as a millimeter wave photon. So like a small steel ball in comparison to a table tennis ping-pong ball, it does not reflect from skin but passes right through the body. This

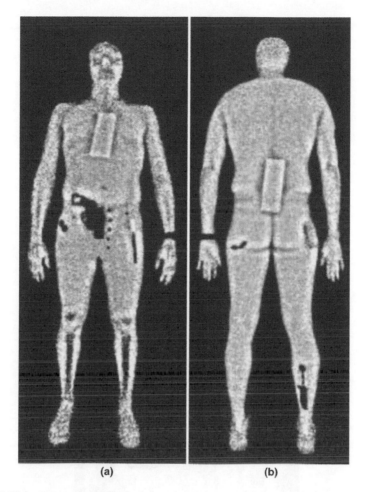

**FIGURE 17.3**  Examples of X-ray of man carrying a concealed gun and packages at his front and back: (a) front view and (b) rear view.

has a slight possibility of internal damage, hence X-ray exposure is considered lifetime accumulative. So unlike millimeter waves, the X-rays cannot be used so easily in backscatter mode. The X-ray dose is very small, so it will take an extremely large number of airport inspections before any problem could be observed for a correctly set up scanner. However, the effects will be added to whatever medical CT and X-ray scans a person has previously been exposed to.

One of the problems with body scanners that see through clothes is public concern about being seen or photographed naked. Some specific concerns are illustrated in the photograph of people protesting in Germany [138] regarding the issue of body piercing and artificial limbs (Figure 17.4). On the girl's chest is written "piercing" and on her right leg "prosthesis."

**FIGURE 17.4**   Protest against full body scanners regarding body piercing and prosthetics.

## 17.3   INSPECTING UNOPENED PACKAGES

Concern has increased in recent years over the shipment of packages with dangerous contents such as bombs, pistols, chemical explosives, and hazardous chemicals. It is generally too expensive to open all packages at sources and destinations. In current X-ray machines for checking contents without opening packages, X-ray particles travel through the package onto a phosphor screen [99]. For a 3D view, the system would require rotation and images would have to be processed with computer tomography (CT). Such machines would be expensive and/or may not provide adequate 3D view of the contents.

We describe a novel millimeter wave approach that has the capability to provide 3D scans cost effectively without opening packages [101]. The approach uses a 94 GHz source that has 3.2 mm wavelength. Previously, we discussed more powerful sources at this frequency (Chapters 10 and 16), but low-power sources for this frequency are also in development [11]. A wavelength of 3.2 mm passes through cardboard and Styrofoam materials, the usual materials for packages. Conducting and other objects in the package will reflect or scatter the radiation. Observation of the scattered radiation in transmission, or backscatter in reflection (Chapter 15), can be analyzed to determine the contents of the package [101]. By tuning the source, it is possible to use a set of images at different frequencies to construct a 3D image of the contents of the package. $K_a$ band radar at 35 GHz can also be used but with a degradation of resolution.

An iterative computation, similar to Gauss–Newton, is described in Section 6.3 and is split into forward and inverse computations.

### 17.3.1   Principles for Proposed Unopened Package Inspection

A handheld scanner passes a short pulse from a collimated beam at 94 GHz through or reflected from the package to be inspected onto a sensor detector array (Figure 17.5). This is somewhat similar to a transmissive or reflective lidar (Chapter 15) and to the W-band radar (Chapter 16). By moving the handheld device across the package, cross-sectional shapes of objects inside the package may be observed. Further, the handheld device can be used from the sides or back of the package to further ascertain the contents. A fixed configuration rather than a handheld device can also be used. The sensor is connected to a laptop computer for displaying the properties of the material along the path of the beam through the package.

*17.3.1.1   Forward Computation*   We use a layered model for propagating the field through the package. The region through which the beam passes in the package is sliced into vertical homogeneous slices (Figure 17.5), each considered to have different homogeneous refractive index and absorption. In layered models, the thicknesses

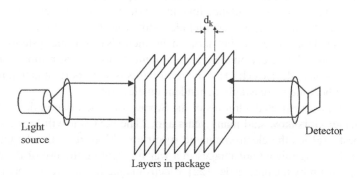

**FIGURE 17.5**   Proposed system for inspection of unopened packages.

of the slices $d_k$ are selected for constant travel time $T$ to simplify the computations [21, 50, 83].

$$d_k = \frac{cT}{n_k} \tag{17.1}$$

Such models are widely used in other field propagation problems for the purpose of computation; geophysics [80, 81] (the latter includes absorption); thin film filter design and testing (Section 6.3); the method of lines in integrated optics; and propagation of light through turbulence (Section 5.4.1). After passing through the package, the light pulse is diffracted through air onto a detector. A matrix method computer algorithm is described in Section 6.3 that estimates the field in each of the vertical slices through which the beam passed. This method was extended to include absorption in Ref. [81]. An output display shows images of the field that passes through or is reflected from the package. Because of multiple scattering between layers, the image will be unclear [104] and further processing will be needed.

**17.3.1.2 *Inverse Computation for Package Inspection*** Because the estimation of parameters for layers is nonlinear, we use the Gauss–Newton iterative method described in Section 6.3.3 to determine the refractive index and absorption [81] of each layer. Starting with a random set of refractive indices in a model, the values are updated iteratively until the output image agrees with that of the model as shown in Figure 6.20. The refractive index and absorption for the layers through which the beam passes or is reflected show the contents of the package. We note that if a short enough source pulse is used so that multiple scattering between layers can be neglected, we can use a faster inverse approach of deconvolution [104].

## 17.4  DESTRUCTION AND PROTECTION OF ELECTRONICS

Because of increasing dependence on electronics in the military, electromagnetic warfare is one of the most critical technologies in the military. We rely on electronics for fire control and on satellites for communications, global positioning (for knowing where everything is), and surveillance imaging to detect conflict preparations (Section 1.3.4.1). Electromagnetic warfare includes both protecting electronics from harm by the enemy and being able to inflict harm on the enemy's electronic systems, including interfering with their electromagnetic communications such as remote control of improvised explosive devices (IEDs). The reason this is so important is that military activities have become increasingly dependent on electronics.

One of the most respected admirals of the Soviet Navy build up in the Cold War, Admiral Sergei Gorshkov, said many years ago, "The next war will be won by the side that best exploits the electromagnetic spectrum." More recently, the Gulf War demonstrated the significant advantages in a war of negating the use of the electromagnetic spectrum by the other side [122]. Electromagnetic warfare is more relevant today than ever before.

**FIGURE 17.6**   W-band tabletop prototype for frying electronics.

## 17.4.1   Interfering or Destroying Enemy Electronics

### 17.4.1.1   *Generating Short Pulses to Destroy Electronics*   Pulses from powerful 94 GHz sources around W-band, described in Section 10.3 and elsewhere in Chapter 10, can be used to interfere with or destroy electronics. For example, a prototype table top system (Figure 17.6) [121] is said to be able to stop an automobile at 100 m. A more powerful system could destroy the electronics at a missile launch site before the missile can take off.

An alternative is the electromagnetic bomb [3, 123] in which an explosive is used to generate an intense electromagnetic field pulse. The electromagnetic bomb uses explosives to move a short circuit into a coil generating a magnetic field so as to compress the **B** field in microseconds [3, 123]. From Maxwell's equation, equation (17.2), this generates a very high electric field pulse **E** around the coil, proportional to the rate of change of the magnetic flux $\partial \mathbf{B}/\partial t$ passing through the coil:

$$\nabla \times \mathbf{E} = -\frac{\partial \mathbf{B}}{\partial t} \tag{17.2}$$

The coil acts as an antenna that radiates the high-intensity electrical field pulse through the air and the pulse induces high voltage and current spikes in electronic circuits to damage them. The waveform, propagating at the speed of light, from a high-altitude source of electromagnetic pulse (EMP) can reach several million square miles of earth.

### 17.4.1.2   *Methods by Which High Peak Power Electromagnetic Pulse Interferes with Electronics*   If the EMP is designed to disrupt communication flow or push a resonator out of synchronization, such as an electronic clock or timing

on a vehicle, a lower power may suffice and the coupling into the system becomes critical. In general, the high peak power field will couple into wires in a similar manner to the way lightning couples into electronics by causing a large voltage spike, especially bad for high-speed low-power electronics, the sort found in military applications where long battery life and fast response matter. The voltage spike finds its way into the electronics. The field can also find its way through gaps in the packaging [136].

### 17.4.2    Protecting Electronics from Electromagnetic Destruction

***17.4.2.1    Nature of Threat***    The increasing dependence of the military on electronics and electrical power connections creates a serious vulnerability to EMP from high-altitude nuclear explosion, an electromagnetic bomb, or an electromagnetic ray gun such as that in Figure 17.6. A crop duster plane flying over a battlefield could drop an electromagnetic bomb to explode in the air above the battlefield to destroy most of the computers, handheld electronic devices, and weapon's electronics within a large area. Hence, in the battlefield, communications, global positioning systems, imaging systems, drones, and weapons electronics such as fire control can fail with disastrous consequences. This is not just a military threat. In assessing the threat of EMP to U.S. electronics, a recent U.S. Government report states, "The damage level could be sufficient to be catastrophic to the Nation, and our current vulnerability invites attack." [51]. In addition to applying all the best practices used in electromagnetic compatibility (EMC) [136], we consider two future approaches.

***17.4.2.2    Replacing Wires with Optical Fibers***    Electromagnetic fields cannot couple into an optical fiber because there are no electrical conductors in a fiber. In fact, during the 1950s, the missile silos constructed in the United States as part of the mutually assured destruction philosophy (MAD) (Chapter 13) for nuclear retaliation in the event of a nuclear attack were reportedly connected to the outside world with optical fiber. This includes electronic and electrical power to electronics. The reason was that a nuclear explosion produces a sequence of high-power electromagnetic pulses that would otherwise couple into electrical wires in the silos and prevent them from functioning. Moreover today, optical fiber technology is more highly developed for the Internet backbone and power fiber has improved to the point where it can replace wire in most electronic and many electrical power applications.

***17.4.2.3    Replacing Electronics with Photons***    Electronic integrated circuits can be replaced with optical integrated circuits [57]. Connections in optical integrated circuits are with optical waveguides and optical fiber is used between chips. Such optical guides have no metal or electronic parts and convey information and power with photons instead of electrons. Photons are uncharged particles, unlike electrons, and are immune to electromagnetic interference.

Over the past 30 years, $300 billion has been spent on electronic very large-scale integration (VLSI), currently approximately $50 billion per year. We can now place 1.9 billion transistors (switching and amplifying elements) on a single half-inch chip. In contrast, today, only a few components are integrated into an optical integrated circuit because integrated optics development started only a few years ago as a result of the optical telecommunications thrust for the Internet. There is currently rapid progress in integrated optics that can be applied to the optics in many military systems.

All-optical computers are immune to EMP and are becoming feasible because of recent developments in the computing and telecommunication industries where optics provides high bit rate for dense communications. Many approaches to constructing an optical computer are described in my book, "Optical Computer Architectures," Wiley, 1991 [83]. However, the technology at that time made all-optical computers expensive, so, as PI, my group in Texas Instruments Central Research Laboratories, designed a hybrid (not all-optical) optical crossbar data flow computer for DARPA [107, 114–116] that was subsequently constructed by Texas Instruments for DoD and is described in Chapter 14 of my book [83]. My recent research on optical computers uses integrated optics for a crossbar switch [102] for a military all-optical computer that is immune to electromagnetic radiation including nuclear blasts and electromagnetic bombs. We need all-optical computers for immunity to electromagnetic attack, to reduce transitions between electronics and optics in telecommunications, to combine optics and electronics on the same silicon chip in large telecommunication routers, to connect more processors on a single chip, and to interface with future faster microprocessors to reduce heat.

# BIBLIOGRAPHY

[1] E. Adams. Flying laser gun. *Popular Science*, 2008.

[2] G. P. Agrawal. *Fiber-Optic Communication Systems*, 3rd edition. Wiley Interscience, New York, 2002.

[3] L. L. Altgilbers, M. D. J. Brown, I. Grishnaev, B. M. Novac, I. R. Smith, I. Tkach, and Y. Tkach. *Magnetocumulative Generators*. Springer, New York, 2000.

[4] L. C. Andrews and R. L. Phillips. *Laser Beam Propagation Through Random Media*, 2nd edition. SPIE Press, Bellingham, WA, 2005.

[5] R. Appleby, H. B. Wallace, and B. Wallace. Standoff detection of weapons and contraband in the 100 GHz to 1 THz region. *IEEE Transactions on Antennas and Propagation*, 55(11):2944–2956, 2007.

[6] F. Bachman, P. Loosen, and R. Proprawe, editors. *High Power Diode Lasers*. Springer, New York, 2007.

[7] C. A. Balanis. *Advanced Engineering Electromagnetics*. Wiley, New York, 1989.

[8] C. Balfour, R. A. Stuart, and A. I. Al-Shamma'a. Experimental studies of backward waves in an industrial free electron laser system. *Optics Communications*, 236:403–410, 2004.

[9] R. J. Barker and E. Schamiloglu, editors. *High-Power Microwave Sources and Technologies*. IEEE Press, New York, 2001.

[10] D. Beason. *The E-Bomb: How America's New Directed Energy Weapons Will Change the Way Future Wars Will Be Fought*. Da Capo Press, Cambridge, MA, 2005.

[11] M. A. Belkin, F. Capasso, F. Xie, A. Belyanin, M. Fischer, A. Wittmann, and J. Faist. Room temperature terahertz quantum cascade laser source based on intracavity difference-frequency generation. *Physics Letters*, 92(20), 2008.

*Military Laser Technology for Defense: Technology for Revolutionizing 21st Century Warfare*,
First Edition. By Alastair D. McAulay.
© 2011 John Wiley & Sons, Inc. Published 2011 by John Wiley & Sons, Inc.

[12] D. Besnard. The megajoule laser program, ignition at hand. *The European Physical Journal D: Atomic, Molecular, Optical and Plasma Physics*, 44(2):207–213, 2007.

[13] J. A. Bittencourt. *Fundamentals of Plasma Physics*, 3rd edition. Springer, New York, 2004.

[14] A. Bjarklev, J. Broeng, and A. Sanchez Bjarklev. *Photonic Crystal Fibres*. Kluwer Academic Publishers, Boston, MA, 2003.

[15] D. Bodanis. $E = mc^2$. Walker Publishing Company, New York, 2000.

[16] M. Born and E. Wolf. *Principles of Optics*, 7th edition. Cambridge University Press, Cambridge, UK, 1999.

[17] G. E. Box and G. Jenkins. *Time-Series Analysis*. Elsevier, 1976.

[18] L. Brekhovskikh. *Waves in Layered Media*. Academic Press, New York, 1980.

[19] T. I. Bunn. Method and system for producing singlet delta oxygen (SDO) and laser system incorporating an SDO generator. U.S. Patent 7,512,169, 2009.

[20] D. K. Cheng. *Fundamentals of Engineering Electromagnetics*. Prentice-Hall, Upper Saddle River, NJ, 1993.

[21] J. F. Claerbout. *Fundamentals of Geophysical Data Processing*. McGraw-Hill, New York, 1976.

[22] C. Von Clausewitz. *On War*. Oxford University Press, New York, 1976. Translated from original 1832.

[23] W. A. Coles, J. P. Filice, R. G. Frehlich, and M. Yadlowsky. Simulation of wave propagation in three-dimensional random media. *Applied Optics*, 34(12):2089–2101, 1995.

[24] R. E. Collin. *Foundations for Microwave Engineering*. McGraw-Hill, New York, 1966.

[25] M. J. Connelly. *Semiconductor Optical Amplifiers*. Kluwer Academic Publishers, Boston, MA, 2002.

[26] R. C. Coutinho, D. R. Selviah, and H. D. Griffiths. High-sensitivity detection of narrowband light in a more intense broadband background using coherence interferogram phase. *Journal of Lightwave Technology*, 24(10), 2006.

[27] H. A. Davis, R. D. Fulton, E. G. Sherwood, and T. J. T. Kwan. Enhanced-efficiency, narrow-band gigawatt microwave output of the reditron oscillator. *IEEE Transactions on Plasma Science*, 18(3):611–617, 1990.

[28] W. Demtröder. *Laser Spectroscopy*, 2nd edition. Springer, New York, 1996.

[29] D. Derickson, editor. *Fiber Optic Test and Measurement*. Prentice-Hall, Upper Saddle River, NJ, 1998.

[30] E. Desurvire. *Erbium-Doped Fiber Amplifiers*. Wiley, New York, 1994.

[31] R. A. Dickerson. Chemical oxygen iodine laser gain generator system. U.S. Patent 6,072,820, 2000.

[32] M. J. F. Digonnet, editor. *Rare-Earth-Doped Fiber Lasers and Amplifiers*. Marcel Dekker, New York, 2001.

[33] R. Drori and E. Jerby. Free-electron-laser-type interaction at 1 meter wavelength range. *Nuclear Instruments and Methods in Physics Research A*, 393:284–288, 1997.

[34] R. W. Duffner. *Airborne Laser: Bullets of Light*. Plenum Press, New York, 1997.

[35] Exotic Electro-Optics. Materials for high-power lasers. Private communication, Exotic Electro-Optics is a subsidiary of II-VI Incorporated, San Diego, CA, August 2010.

[36] J.-F. Eloy. *Power Lasers*. Wiley, New York, 1987. Translated from French original 1985 to English in 1987 by H. Beedie.

[37] M. H. Fields, J. E. Kansky, R. D. Stock, D. S. Powers, P. J. Berger, and C. Higgs. Initial results from the advanced-concepts laboratory for adaptive optics and

tracking. In *Proceedings of SPIE, Laser Weapons Technology*, Vol. 4034, April 2000, pp. 116–127.

[38] W. Fitzgerald. A 35 GHz beam waveguide system for millimeter-wave radar. *Lincoln Laboratory Journal*, 5(2):245–272, 1992.

[39] G. E. Forden. The airborne laser. *IEEE Spectrum*, 34(9), pp. 40–49, 1997.

[40] A. V. Gaponov-Grekhov and V. L. Granatstein, editors. *Applications of High-Power Microwaves*. Artech House, Boston, 1994.

[41] J. A. Gaudet, R. J. Barker, C. J. Buchenauer, C. Christodoulou, J. Dickens, M. A. Gundersen, R. P. Joshi, H. G. Krompholz, J. F. Kolb, A. Kuthi, M. Laroussi, A. Neuber, W. Nunnally, E. Schamiloglu, K. H. Schoenbach, J. S. Tyo, and R. J. Vidmar. Research issues in developing compact pulsed power for high peak power applications on mobile platforms. *Proceedings of the IEEE*, 92(7):1144–1165, 2004.

[42] I. M. Gelfand and S. V. Fomin. *Calculus of Variations*. Prentice-Hall, Englewood Cliffs, NJ, 1963.

[43] R. W. Gerchberg and W. O. Saxton. A practical algorithm for the determination of phase from image and diffraction plane pictures. *Optik*, 35(2):237–246, 1972.

[44] A. Gerrard and J. M. Burch. *Introduction to Matrix Methods in Optics*. Dover Publications, New York, 1975.

[45] A. K. Ghatak, K. Thyagarajan, and M. R. Shenoi. Numerical analysis of planar optical waveguides using matrix approach. *Journal of Lightwave Technology*, 5(5):660–667, 1987.

[46] D. C. Ghiglia and M. D. Pritt. *Two-Dimensional Phase Unwrapping*. Wiley, New York, 1998.

[47] C. Gomez-Reino, M. V. Perez, C. Bao, and V. Perez. *Gradient-Index Optics*. Springer, New York, 2002.

[48] J. W. Goodman. *Statistical Optics*. Wiley, New York, 1985.

[49] J. W. Goodman. *Introduction to Fourier Optics*, 3rd edition. Roberts and Company Publishers, Englewood, CO, 2005. Previously McGraw-Hill, New York, 1996.

[50] P. Goupillaud. An approach to inverse filtering of near surface layer effects from seismic records. *Geophysics*, 26(6):754–760, 1961.

[51] U.S. Government. Report of the Commission to Assess the Threat to the United States from Electromagnetic Pulse (EMP) Attack. http://www.globalsecurity.org/wmd/library/congress/2004_r/04-07-22emp.pdf, 2004.

[52] J. G. Graeme. *Photodiode Amplifiers*. McGraw-Hill, New York, 1996.

[53] P. E. Green. *Fiber Optic Networks*. Prentice-Hall, Englewood Cliffs, NJ, 1993.

[54] C. M. Harding, R. A. Johnston, and R. G. Lane. Fast simulation of a Kolmogorov phase screen. *Applied Optics*, 38(11):2161–2170, 1999.

[55] P. Harihanran. *Basics of Interferometry*. Academic Press, San Diego, CA, 1992.

[56] O. S. Heavens. *Optical Properties of Thin Solid Films*. Dover Publications, 1991. Original publication 1955.

[57] R. G. Hunsperger. *Integrated Optics*, 5th edition. Springer, New York, 2002.

[58] Google images of ABL. Images of ABL. http://www.google.com/images?rlz=1T4ACAW_enUS341US346&q=airborne+laser+photograph&um=1&ie=UTF-8&source=univ&ei=WwpoTKncOYP7lweV_OmgBQ&sa=X&oi=image_result_group&ct=title&resnum=1&ved=0CCYQsAQwAA, 2010.

[59] Google images of BAE Jeteye. Images of BAE JetEye program. http://www.baesystems.com/BAEProd/groups/public/documents/bae_publication/bae_pdf_eis_jeteye.pdf, 2010.

[60] A. Ishimaru. *Wave Propagation and Scattering in Random Media*. IEEE Press, Piscataway, NJ, 1997. Previously published by Academic Press in Vols. 1 and 2 in 1978.

[61] F. A. Jenkins and H. E. White. *Fundamentals of Optics*, 4th edition. McGraw-Hill, New York, 1976.

[62] D. Jensen. Missile defense: post DHS. http://www.militaryaerospace.com/index/display/avi-article-display.372267.articles.avionics-intelligence.news.2010.01.missile-defense-post-dhs.html, 2010.

[63] J. D. Joannopoulos, R. D. Meade, and J. N. Winn. *Photonic Crystals: Molding the Flow of Light*. Princeton University Press, Princeton, NJ, 1995.

[64] L. Johnson, F. Leonberger, and G. Pratt. Integrated-optic temperature sensor. *Applied Physics Letters*, 41(2):134–136, 1982.

[65] M. V. Kartikeyan, E. Borie, and M. K. A. Thumm, *Gyrotrons*. Springer, New York, 2004.

[66] L. A. Klein. *Millimeter-Wave and Infrared Multisensor Design and Signal Processing*. Artech House, Boston, MA, 1997.

[67] W. Koechner. *Solid-State Laser Engineering*, 3rd edition. Springer, New York, 1992.

[68] A. N. Kolmogorov. The local structure of turbulence in an incompressible viscous fluid for very large Reynolds numbers. *Comptes Rendus (Doklady) Academy of Sciences, USSR*, 30:301–305, 1941.

[69] B. Kress and P. Meyrueis. *Digital Diffractive Optics*. Wiley, New York, 2000.

[70] T. J. T. Kwan and C. M. Snell. Virtual cathode microwave generator having annular anode slit. U.S. Patent 4,730,170, March 1988. Assigned to U.S. Department of Energy.

[71] M. B. Lara, J. Mankowski, J. Dickens, and M. Kristiansen. Reflex-triode geometry of the virtual-cathode oscillator. In *Digest of Technical Papers, 14th IEEE International Pulsed Power Conference*, Vol. 2, June 2003, pp. 1161–1164.

[72] W. Lauterborn and T. Kurz. *Coherent Optics*, 2nd edition. Springer, New York, 2003.

[73] S. Liao. Miter bend design for corrugated waveguides. *Progress in Electromagnetics Research Letters*, 10:157–162, 2009. http://ceta.mit.edu/PIERL/pierl10/17.09062103.pdf.

[74] G. J. Linde, M. T. Ngo, B. G. Danly, W. J. Cheung, and V. Gregers-Hansen. WARLOC: a high-power coherent 94 GHz radar. *IEEE Transactions on Aerospace and Electronic Systems*, 44(3):1102–1107, 2008.

[75] M. M.-K. Liu. *Principles and Applications of Optical Communications*. Irwin, Boston, 1996.

[76] W. M. Manheimer. On the possibility of high power gyrotrons for super range resolution radar and atmospheric sensing. *International Journal of Electronics*, 72(6):1165–1189, 1992.

[77] J. M. Martin and S. M. Flatte. Intensity images and statistics from numerical simulation of wave propagation in 3-D random media. *Applied Optics*, 27(11):2111–2126, 1988.

[78] A. D. McAulay. The finite element solution of dissipative electromagnetic surface waveguides. *International Journal for Numerical Methods in Engineering*, 11(1):11–27, 1977.

[79] A. D. McAulay. Variational finite element solution of dissipative waveguides and transportation application. *IEEE Transactions on Microwave Theory and Techniques*, 25(5):382–392, 1977.

[80] A. D. McAulay. Prestack inversion with plane-layer point source modeling. *Geophysics*, 50(1):77–89, 1985.

[81] A. D. McAulay. Plane-layer prestack inversion in the presence of surface reverberation. *Geophysics*, 51(9):1789–1800, 1986.

[82] A. D. McAulay. Engineering design neural networks using split inversion learning. In *Proceedings of the IEEE First International Conference on Neural Networks*, Vol. IV, June 1987, pp. 635–641.

[83] A. D. McAulay. *Optical Computer Architectures*. Wiley, New York, 1991.

[84] A. D. McAulay. Modeling of deterministic chaotic noise to improve target recognition. In *Proceedings of SPIE, Signal Processing, Sensor Fusion, and Target Recognition Conference*, Vol. 1955-17, April 1993, pp. 50–57.

[85] A. D. McAulay. Optical recognition of defective pins on VLSI chips using electron trapping material. In *Proceedings of SPIE, Optical Implementation of Information Processing Conference*, July 1995.

[86] A. D. McAulay. Diffractive optical element for multiresolution preprocessing for computer vision. In *Proceedings of SPIE, Signal Processing, Sensor Fusion, and Target Recognition Conference VIII*, Vol. 3720-43, April 1999.

[87] A. D. McAulay. Improving bandwidth for line-of-sight optical wireless in turbulent air by using phase conjugation. In *Proceedings of SPIE, Optical Wireless Communications II*, Vol. 3850-5, September 1999.

[88] A. D. McAulay. Generating Kolmogorov phase screens for modeling optical turbulence. In *Proceedings of SPIE, Laser Weapons Technology*, Vol. 4034-7, April 2000.

[89] A. D. McAulay. Optical arithmetic unit using bit-WDM. *Optics and Laser Technology*, 32:421–427, 2000.

[90] A. D. McAulay. Artificial turbulence generation alternatives for use in computer and laboratory experiments. In *Proceedings of SPIE, High Resolution Wavefront Control: Methods, Devices, and Applications III*, Vol. 4493, August 2001, pp. 141–149.

[91] A. D. McAulay. All-optical SOA latch fail-safe alarm system. In *Proceedings of SPIE, Photonic Devices and Algorithms for Computing VI*, Vol. 5556-7, August 2004, pp. 68–72.

[92] A. D. McAulay. Novel all-optical flip-flop using semiconductor optical amplifiers in innovative frequency-shifting inverse-threshold pairs. *Optical Engineering*, 45(5):1115–1120, 2004.

[93] A. D. McAulay. Optical bit-serial computing. *Encyclopedia of Modern Optics*, 2004.

[94] A. D. McAulay. Optimizing SOA frequency conversion with discriminant filter. In *Proceedings of SPIE, Active and Passive Optical Components for WDM Communications IV*, Vol. 5595, October 2004, pp. 323–327.

[95] A. D. McAulay. Leaky wave interconnections between integrated optic waveguides. In *Proceedings of SPIE, Active and Passive Optical Components for WDM Communications V*, Vol. 6014-17, October 2005, pp. OC-1–OC-8.

[96] A. D. McAulay. Nonlinear microring resonators forge all-optical switch. *Laser Focus World*, November 2005, pp. 127–130.

[97] A. D. McAulay. Computing fields in a cylindrically curved dielectric layered media. In *Proceedings of SPIE, Enabling Photonic Technologies for Defense, Security and Aerospace Applications VII*, Vol. 6243-18, April 2006, pp. 1–8.

[98] A. D. McAulay. Modeling the brain with laser diodes. In *Proceedings of SPIE, Active and Passive Optical Components for Communications VII*, Vol. 6775-10, October 2007.

[99] A. D. McAulay. Novel lock-in amplifier for identification of luminescent materials for authentication. In *Proceedings of SPIE, Signal Processing, Sensor Fusion, and Target Recognition XVI*, Vol. 6567-48, April 2007.

[100] A. D. McAulay. Frustrated polarization fiber Sagnac interferometer displacement sensor. In *Proceedings of SPIE, Signal Processing, Sensor Fusion, and Target Recognition XVII*, Vol. 6969-50, March 2008.

[101] A. D. McAulay. Package inspection using inverse diffraction. In *SPIE Optics and Photonics for Information Processing II*, Vol. 7072-16, August 2008.

[102] A. D. McAulay. Digital crossbar switch using nonlinear optical ring resonator. In *Proceedings of SPIE, Optics and Photonics for Information Processing III*, Vol. 7442-2, August 2009.

[103] A. D. McAulay. Integrated optic chip for laser threat identification. In *Proceedings of SPIE, Signal Processing, Sensor Fusion, and Target Recognition XIX*, Vol. 7697-48, April 2010.

[104] A. D. McAulay. Optical deconvolution for multilayer reflected data. In *Proceedings of SPIE, Optics and Photonics for Information Processing IV*, Vol. 7797-8, August 2010.

[105] A. D. McAulay, M. R. Corcoran, C. J. Florio, and I. B. Murray. Optical micro-ring resonator filter trade-offs. In *Proceedings of SPIE, Active and Passive Optical Components for WDM Communications IV*, Vol. 5595-48, October 2004, pp. 359–364.

[106] A. D. McAulay, M. R. Corcoran, C. J. Florio, and I. B. Murray. All optical switching and logic with an integrated optic microring resonator. In *Proceedings of SPIE, Enabling Photonic Technologies for Defense, Security and Aerospace Applications VI*, Vol. 5814-3, March 2005, pp. 16–22.

[107] A. D. McAulay, D. W. Oxley, R. W. Cohn, J. D. Provence, E. Parsons, and D. Casasent. Optical crossbar interconnected signal processor. Report, DARPA/ONR N00014-85-C-0755, 1991.

[108] A. D. McAulay, K. Saruhan, and A. Coker. Real-time computation for absorption free beamforming using neural networks. In *Proceedings of the IEEE International Conference on Systems Engineering*, August 1990.

[109] A. D. McAulay and H. Tong. Modeling neural networks with active optical devices. In *Proceedings of SPIE, Optical Information Systems III*, Vol. 5908-15, August 2005.

[110] A. D. McAulay and H. Tong. Optical clustering for unsupervised learning using coupled microring resonators. In *Proceedings of SPIE, Signal Processing, Sensor Fusion, and Target Recognition XIV*, Vol. 5809-49, April 2005, pp. 402–408.

[111] A. D. McAulay and J. Wang. Optical diffraction inspection of periodic structures using neural networks. *Optical Engineering*, 37(3):884–888, 1998.

[112] A. D. McAulay and J. Wang. A Sagnac interferometer sensor system for intrusion detection and localization. In *Proceedings of SPIE, Enabling Photonic Technologies for Aerospace Applications V*, Vol. 5435-16, April 2004, pp. 114–119.

[113] A. D. McAulay. Deformable mirror nearest neighbor optical computer. *Optical Engineering*, 25(1):76–81, January 1986.

[114] A. D. McAulay. Optical crossbar interconnected signal processor with basic algorithms. *Optical Engineering*, 25(1):82–90, 1986.

[115] A. D. McAulay. Spatial light modulator interconnected computers. *Computer*, 20(10):45–57, 1987.

[116] A. D. McAulay. Conjugate gradients on optical crossbar interconnected multiprocessor. *Journal of Parallel and Distributed Processing*, 6:136–150, 1989.

[117] Megajoule. French Megajoule Laser Facility. http://en.wikipedia.org/wiki/Laser_M%C3%A9gajoule, 2009.

[118] A. Mendez and T. F. Morse. *Specialty Optical Fibers Handbook*. Elsevier/Academic Press, Boston, MA, 2007.

[119] P. W. Milonni and J. H. Eberly. *Lasers*. Wiley, New York, 1988.

[120] A. W. Miziolek, V. Palleschi, and I. Schecter, editors. *Laser-Induced Breakdown Spectroscopy*. Cambridge University Press, Cambridge, UK, 2006.

[121] G. Murdoch. Blackout bomb: Air Force's high-powered microwave weapons fry enemy equipment. *Popular Science*, 2009.

[122] V. K. Nair. *War in the Gulf: Lessons for the Third World*. Lancer International, New Delhi, 1991.

[123] A. A. Neuber, editor. *Explosively Driven Pulse Power: Helical Magnetic Flux Compression Generators*. Springer, New York, 2005.

[124] M. J. Neufeld. *Von Braun*. A. A. Knopf, New York, 2007.

[125] PR News. American Society for Photogrammetry and Remote Sensing web site, guide to land imaging satellites. http://www.asprs.org/news/satellites/, 2009.

[126] PR News. Raytheon awarded contract for Office of Naval Research's free electron laser program. PR News, June 9, 2009. http://www.prnewswire.com/comp/149999.htm, 2009.

[127] NIF. National Ignition Facility, or NIF. http://en.wikipedia.org/wiki/National_Ignition_Facility, 2009.

[128] Northrop-Grumman. Northrop-Grumman Guardian images. http://www.es.northropgrumman.com/countermanpads/media_gallery/photos.html, 2010.

[129] G. S. Nusinovich. *Introduction to the Physics of Gyrotrons*. Johns Hopkins University Press, Baltimore, MD, 2004.

[130] K. Okamoto. *Fundamentals of Optical Waveguides*. Academic Press, New York, 2000.

[131] T. Okoshi and K. Kikuchi. *Coherent Optical Fiber Communications*. KTK Scientific Publishers/Kluwer Academic Publishers, Tokyo/Boston, MA, 1988.

[132] E. L. O'Neill. *Introduction to Statistical Optics*. Dover Publications, New York, 1991.

[133] A. V. Oppenheim and R. W. Schafer. *Discrete-Time Signal Processing*, 2nd edition. Prentice-Hall, Upper Saddle River, New Jersey, 1999.

[134] A. Papoulis. *Systems and Transforms with Applications in Optics*. R. E. Krieger Publishing Company, 1981. Original printing 1968.

[135] A. Papoulis. *Probability, Random Variables, and Stochastic Processes*, 3rd edition. McGraw-Hill, New York, 1991.

[136] C. R. Paul. *Introduction to Electromagnetic Compatibility*. Wiley, New York, 1992.

[137] J. Pearson. The Electrodynamic Debris Eliminator (EDDE): removing debris from space. *The Bent of Tau Beta Pi*, C1(2), 2010.

[138] M. Phillips. From granny to nearly nude germans, Everyone's raising cane at the airport. *Wall Street Journal*, January 11, 2010.

[139] Active Denial Program. Active Denial System on Humvee. http://en.wikipedia.org/wiki/File:Active_Denial_System_Humvee.jpg, 2010.

[140] R. Rhodes. *Arsenals of Folly*. A. A. Knopf, New York, 2007.

[141] M. A. Richards. *Fundamentals of Radar Signal Processing*. McGraw-Hill, New York, 2005.

[142] R. D. Richmond and S. C. Cain. *Direct-Detection LADAR Systems*. SPIE Publications, Bellingham, WA, 2010.

[143] J. Robieux. *High Power Laser Interactions*. Lavoisier Publishing, Secaucus, NJ, 2000.

[144] M. C. Roggemann and B. Welsh. *Imaging Through Turbulence*. CRC Press, New York, 1996.

[145] N. Roy and F. Reid. Off-axis laser detection model in coastal areas. *Optical Engineering*, 47(8), 2008.

[146] B. S. Woodard, J. W. Zimmerman, G. F. Benavides, D. L. Carroll, J. T. Verdeyen, A. D. Palla, T. H. Field, W. C. Solomon, S. J. Davis, W. T. Rawlins, and S. Lee. Gain and continuous-wave laser oscillation on the 1315 nm atomic iodine transition pumped by an air–helium electric discharge. *Applied Physics Letters*, 93:021104, 2008.

[147] K. Sakoda. *Optical Properties of Photonic Crystals*. Springer, New York, 2001.

[148] B. E. A. Saleh and M. C. Teich. *Fundamentals of Photonics*. Wiley, 1991.

[149] M. A. Shapiro and R. J. Temkin. High power miter-bend for the next linear collider. In *Proceedings of the 1999 Particle Accelerator Conference*, 1999.

[150] D. M. Sheen, D. L. McMakin, H. D. Collins, T. E. Hall, and R. H. Stevertson. Concealed explosive detection on personnel using a wideband holographic millimeter-wave imaging system. In *Proceedings of SPIE, Signal Processing, Sensor Fusion, and Target Recognition V*, Vol. 2755, 1996.

[151] M. Skolnik. *Radar Handbook*, 3rd edition. McGraw-Hill, New York, 2008.

[152] V. A. Soifer, editor. *Methods for Computer Design of Diffractive Optical Elements*. Wiley Interscience, 2002.

[153] S. H. Strogatz. *Nonlinear Dynamics and Chaos*. Addison-Wesley, Reading, MA, 1994.

[154] A. Taflove and S. C. Hagness. *Computational Electrodynamics: The Finite Difference Time-Domain Method*, 2nd edition. Artech House, Boston, MA, 2000.

[155] V. I. Tatarski. *Wave Propagation in a Turbulent Medium*. McGraw-Hill, New York, 1961. Translated from Russian by R. A. Silverman.

[156] K. Thayagaran, M. R. Shenoi, and A. K. Ghatak. Accurate method for the calculation of bending loss in optical waveguides using a matrix approach. *Optics Letters*, 12(4):296–298, 1987.

[157] M. K. A. Thumm. State of the art of high power gyro devices and free electron masers. Technical Report FZKA 7198, Institute fur Hochleistungsimplus- und Mikrowellen-technik, Karlsruhe FZKA, 2008.

[158] A. Tolkachev. Gyroklystron-based 35 GHz radar for observation of space objects. In *Proceedings of the 22nd International Conference on Infrared and Millimeter Waves*, Vol. 5595-48, October 1997, pp. 359–364.

[159] A. Tolkachev, V. Trushin, and V. Veitsel. On the possibility of using powerful millimeter wave band radars for tracking objects in circumterrestrial space. In *Proceedings of the 1996 CIE International Conference of Radar*, 1996.

[160] H. Tong and A. D. McAulay. Wavefront measurement by using photonic crystals. In *Proceedings of SPIE, Enabling Photonic Technologies for Aerospace Applications VI*, Vol. 5435-13, April 2004.

[161] H. L. Van Trees. *Detection, Estimation and Modulation Theory, Part I*. Wiley, 2001. Originally published in 1968.

[162] H. L. Van Trees. *Detection, Estimation and Modulation Theory, Part IV. Optimum Array Processing*. Wiley Interscience, 2002.

[163] R. K. Tyson. *Principles of Adaptive Optics*, 2nd edition. Academic Press, New York, 1998.

[164] E. Udd, editor. *Fiber Optic Sensors*. Wiley, New York, 1991.

[165] J. Vetrovec. Chemical oxygen–iodine laser (coil)/cryosorption vacuum pump system, 2000.

[166] U. Wandinger. Introduction to lidar. In C. Weitkamp, editor, *Lidar Range-Resolved Optical Remote Sensing of the Atmosphere*. Springer, New York, 2005, Chapter 1.

[167] B. Warm, D. Wittner, and M. Noll. Apparatus for defending against an attacking missile. U.S. Patent 5,600,434, 1997.

[168] C. Weitkamp, editor. *Lidar Range-Resolved Optical Remote Sensing of the Atmosphere*. Springer, New York, 2005.

[169] H. Weyl. *Theory of Groups and Quantum Mechanics*. Dover Publications, 1950.

[170] R. Whitney, D. Douglas, and G. Neil. Airborne megawatt class free-electron laser for defense and security. In *2005 Conference: Laser Source and System Technology for Defense and Security*, Jefferson Laboratory, 2008.

[171] S. Wieczorek, B. Krauskopf, and D. Lenstra. A unifying view of bifurcations in a semiconductor laser subject to optical injection. *Optics Communications*, 172.1:279–295, 1999.

[172] P. P. Woskov, V. S. Bajaj, M. K. Hornstein, R. J. Temkin, and R. G. Griffin. Corrugated waveguide and directional coupler for CW 250-GHz gyrotron DNP experiments. *IEEE Transactions on Microwave Theory and Techniques*, 53(6):1863–1869, 2005.

[173] C. C. Wright, R. A. Stuart, A. Al-Shamma'a, and A. Shaw. Free electron maser as a frequency agile microwave source. In *Loughborough Antennas and Propagation Conference*, April 2007, pp. 325–328.

[174] C. C. Wright, R. A. Stuart, J. Lucas, and A. Al-Shamma'a. Low cost undulator magnets for industrial free electron masers. *Optics Communications*, 185:387–391, 2000.

[175] C. C. Wright, R. A. Stuart, J. Lucas, A. Al-Shamma'a, and C. Petichakis. Design and construction of a table top microwave free electron maser for industrial applications. *Vacuum*, 77:527–531, 2005.

[176] A. Yariv and P. Yeh. *Photonics: Optical Electronics in Modern Communications*, 6th edition. Oxford University Press, Oxford UK, 2007.

[177] J. L. Zhang, E. M. Tian, and Z. B. Wang. Research on coherent laser warning receiver based on sinusoidal transmission grating diffraction. *Journal of Physics: Conference Series*, 48, 2006.

# INDEX

*Military Laser Technology for Defense: Technology for Revolutionizing 21st Century Warfare*,
First Edition. By Alastair D. McAulay.
© 2011 John Wiley & Sons, Inc. Published 2011 by John Wiley & Sons, Inc.

Printed and bound by CPI Group (UK) Ltd, Croydon, CR0 4YY

27/10/2024

14580133-0004